Anti-Vigilante and the Rips in Time

By Diane Tegarden

Library of Congress Cataloging-in-Publication Data

Tegarden, Diane 1957-
Anti-Vigilante and The Rips in Time/ Diane Tegarden.

ISBN 0-9745369-1-1

1. Science Fiction
2. Renewable Energy
3. Hopitu Shinumu Culture
4. Ocean Energy

FireWalker Publications, Inc
3579 E. Foothill Blvd. #322
Pasadena, CA 91107-3119

Interior Book Design: Cindy Clark of CCDesignz Business Services, LLC
Cover Art Design: Kristine Sheehan of The Merry Bird

Printed in the United States by Lightning Source, Inc.

This is a "preventable vision of the future", if people care enough about our planet, and act in time, we'll still have a beautiful place to live.... if not, we are doomed by our own obstinacy to live in this future world.

Your choice,
Diane Tegarden

CONTENTS

Acknowledgements .... *vii*

CHAPTER 1
The Rips in Time 1

CHAPTER 2
Who Rules the M.O.B.? 71

CHAPTER 3
Anti-Vigilante is born 141

CHAPTER 4
The AGGIE Meet-up 199

CHAPTER 5
Cami, Chameleon 257

CHAPTER 6
The Dark Ones 297

Rave Reviews *353*

Sources Cited *355*

About the Author *356*

Acknowledgements

This book was a labor of love that took me three years to write, but I certainly couldn't have done it without help.

I'd like to thank my brother Phoenix Valencia who read the first chapter and wowed me with his comments and encouragement. Dennis "Docc" Rudolph who did the preliminary editing on chapters 1-6 and even got into an argument with one of my characters; talk about getting emotionally involved with the material!

I'd like to recognize the mentoring genius of Anthony Robbins who helped me to believe in myself and never accept less than the best effort out of my work.

My beloved husband, Wade Webb, for his solid encouragement, love and understanding that it takes to live with someone whose mind is a billion miles away.

Special thanks and kudos go out to my great friend, Cindy Clark who designed the interior of the book and helped with editing, I couldn't have finished this book without her attention to detail, endless patience and unbounded enthusiasm for my work.

I also want to recognize the talent of my covert art designer, Kristine Sheehan, who came up with the most intriguing cover art, making my book stand out in a massive sci-fi market.

1

The Rips in Time

Debt of Pain

"I dared to speak against your will,
I let another truth be known.
And in that moment you hated me,
Your essence, with that hate, doth shone."

- FROM DEBT OF PAIN, BY THE BARD PHOENIX GAET

The Milliner

The Milliner did not know what type of creature he was, he only knew that his overwhelming predisposition for violence absolutely guaranteed he would commit unspeakable acts without so much as the bat of an eyelash. His pale washed out icy blue eyes looked through rather than at his victims.

He lived in a chronic, cushy pad at the top of the Hive. It was lavishly under- furnished in the meanest lam-plas and fer-rocement decor available, his license to jack-in almost unlimited. He stayed in this spacious room playing with his antique reelies, grammees and nanotech toys, awaiting his next work assignment as casually as one would sort spam. When the big bosses blew up his celly he would tune in immediately, even if he was accessing his favorite Senselle program.

He considered it a mental discipline to viciously rip out his jack-in, mid-session. Usually used as a form of suicide, this wasn't a common practice, but the Milliner had schooled himself to neither enjoy nor deny pain of any sort, whether it was his or another's. That was the secret of keeping his near royalty status among the BayBays, they tuned-in to his 'gramming for his one unique talent, his ability to feed a chaotic strain of unplanned violence to his greedy viewership. They could never get enough of it.

He didn't know and didn't care who he worked for; whether it was the Senselles, (who controlled restrictions and enhancements to the genetic engineering licensing for all BayBays); the MOB, with their stranglehold on all electronic entertainment media; or the idiot-savant MagicoMathematicians, with their formulas and fright wigs. You see, he was abundantly recompensed by the shadowy corporation who employed him. Although he would never know who employed him, he didn't care as long as the pay was good. That's all that mattered.

Besides, the tiny, nagging mystery of who they were gave his mind something pleasurable to toy with, since his victims had long since become easy meat.

He returned to his latest fix from the Sextel Channel, featuring Seleste Boobastis, informally known as the Cat Lady.

As he tuned-in he heard the familiar voice of a Talking Head intoning....

We now return you to "A Seleste Boobastis Programme"...

* * *

This began like most mornings, Seleste and her husband, Ray, had had an argument. It was about his endless preoccupation with his work. All he did was work. He used most of the day to work and was so tired by the time he got home that he had little energy for Seleste or the kids. She had had it with his endless excuses, telling her she had to "wait her turn" for his time and affection. After they had bitterly argued about it for the millionth time, he broke the news that he was going to get yet another part time job to supplement their income.

Seleste lost it and started crying, something Ray hated.

"You just don't understand." She knew she sounded pathetic and needy, but couldn't help herself and went on.

"I have to beg you for any little scrap of time and all you can do is distance me even further."

After a long, blank-eyed moment passed, he looked her with thin pity, shook his head and turned away. As he ambled out the door, his thoughts turned to the job he was doing today, and he didn't give her another thought.

Seleste knew the argument was over for the moment and wished fervently she had never married the schmuck, but here she was, stuck in a miasma of alcoholism and depression. She couldn't find the strength to go through another angry divorce, so just tried to cope until she could figure out what to do.

Seleste was thoroughly sick of Raymond, tired of the way he was always cleaning up after her, after she had just cleaned the house.

Say she would clean the kitchen counters, then "the idiot" would come along after her with bleach or something and clean up after her! Like she was a slob or something. It would drive her absolutely bats. Nothing was ever quite good enough for him no matter how often she'd clean the floor, or vacuumed up after the kitties; he'd always complain about cat dander everywhere and he would wonder loudly why on earth he ever wanted her in the first place.

When they hooked up, Raymond knew that Seleste had 14 cats – that she loved them, and would never give them up for him. Seleste adored all their feline idiosyncrasies, because cats had been emotionally supportive of her in times of loneliness and sorrow. Cats had always comforted her when no one else in her life was there for her.

Her slowly spiraling thoughts returned her to her little charges. Thyla, a little orange, white and black spattered kitten was playing with her mother, Osmo (short for Osmosis). Seleste remembered when she had first set eyes on Osmo, long ago.

She had found her wandering, lost and frightened in the back alley of her apartment. Seleste had walked out to the back to take her trash out to the big dumpster and she saw Osmo standing there, piteously crying. Osmo was so skinny you could see every bone in her body and her voice cracked from exhaustion. The kitty was on her last legs, you could see it.

Seleste scooped her up with one hand and cuddled the kitten to her chest, immediately falling in love with her. Osmosis was a special little lady. Her eyes were blue – crackling, bright blue- and she had the most inquisitive mind. Seleste would come to know all this in time, but for the moment the kitten had ear mites; was flea bitten; starving and badly dehydrated.

Bundling the kitten into an old sweater, Seleste jumped into her beat up old Volvo, (the one that her friends told her was an indecipherable color – not quite a gun metal grey, something closer to. . . primer), and caromed down the road to her vets office.

After all the requisite shots that were allowed for a kitten that age (she was only four or five weeks old and shouldn't have been away from her mom, but obviously, had been abandoned), Seleste took Osmo home and they clicked immediately.

Osmo was one of the first kitties to ride on Seleste's shoulder and since Os never really got that big, it was no imposition at all. Osmo would sit on Seleste's right shoulder and purr and click in her ear, as though having a personal conversation with her. If Seleste had been inclined to be a witch, she might have thought of Osmo as her "familiar". Being open minded about god, but basically an atheist she didn't adhere to such nonsense, although she was intellectually interested in witchcraft and things of that sort, and would discuss it endlessly with her old friend, Echo. Still, Seleste didn't really ascribe to any of it personally.

No matter. This morning, as Seleste watched the sun slowly creep up the orangish-red sky, she took a deep breath of the mingled scents of Eucalyptus, Dusty Oak, and Norfolk Pines from the surrounding yard. She also had fruit trees, although they weren't in bloom this time of year, so they didn't contribute to the floral bouquet.

It was the beginning of summer and Seleste knew it was going to be a hot, dry day by the stillness in the air. The insects that were usually clicking and buzzing outside since the early a.m. had already settled down. They were no longer active and neither were the kitties, who had all decided that it was a good time to take their early morning snooze.

Bappas was casually laying on the ratty old tan colored couch that was backed against the living room window facing east. The ledge there provided just enough sun for Bappas – a fat, lazy Tom cat of about nine years, who was painted with the distinctive solid orange stripes of a tiger.

His stripes were clearly defined, and he was proud to display them as he lay about on the window sill and let the morning sun warm his old bones. He was looking out the window, watching the younger ones at play as they went tumbling around and swiped at each other. Trios of them merrily jumping straight up in the air, and then scrambling off in three directions like lunatic cartoon characters. But they didn't interest him quite as much as getting himself into just the right position, so that the sun could warm his whole back. And once set, he was usually good for the whole morning.

Hell, for the whole day.

Seleste wondered where her Little Bit was – he was the newcomer to the bunch. A curious two year old tabby cat she had adopted as a tiny kitten, he was always stirring the others to new mischief.

She went over to look out the window for him and got lost in her own thoughts. Thoughts swirled inside her head like pleasant dreams, while still being awake. She was thinking of other places she had been, other times she had lived in, and other men that she had loved.

As she pondered her past Sir Handsome, a pure black, satiny coated alley cat, came over to her and silently leapt onto the windowsill she was leaning against. He positioned himself between the window and her body, so he would fit perfectly against her at the hip level, and she instinctually put her arms around him. Then she absently wandered from window to window, room to room of her house with him in her arms. She didn't know why she did this ritual, but they both enjoyed it nonetheless.

She would look out in each direction. In her mind, memorizing the position of each tree, each bush and flower she grew. She observed the vibrancy of their colors change as the sun was peeking over the tops of the trees in front of her house. Then slowly, in streamers, the sun would pierce through leaves, foliage, and flowers, illuminating them as if alit from inside, their colors becoming vibrant, completely unlike how they looked when they were shaded. The dull grey-greens turned to deep forest greens and lively Kelly greens and delicate lime greens. The leaves would reflect and intensify in their color, like a painting coming to life.

Putting down Sir Handsome, Seleste slowly wandered back to where she had left her coffee, cupping the bottom with her left hand and tracing the decorations on the outside of the cup with her right. Slowly bringing it to her lips, she sipped the hot bitter fluid. It tasted wonderful.

* * *

The Milliner's jack-in ended abruptly, he was being peeped.

His employers being as quixotic as ever, IM'd the new assignment directly into his frontal lobe, couched as a subtle suggestion that he take a little trip DownTown to infiltrate a meeting of the AGGIEs. He is given lat's and lon's to the meeting place and given explicit instructions to use the very public slidewalks to get to the nearest Dissemblage Portal.

Without so much as a glance back at his cubby, he flicked a switch on his custom tailored inky black bodysuit and changed the crimson red scorpion hologram he was wearing to that of a top-of-the-line Alfa Romeo replica. Looking at his reflection in his vidscreen, he added a magenta Targa helmet and jumped into his fastest Lamborghini jetboots with the mini-flamethrowers. He knew this was all the rage with the lowlifer techies below, and rather than look preposterous, he would easily blend in.

Taking a public slidewalk to the nearest DP, he felt the familiar urge to cause havoc immediately, tearing into these useless lumps of flesh surrounding him that he considered subhuman. He detested Pseudo-Humanoid Drones although his encounters with them had been few and far between. The DownTowners lived in The Warrens below the surface of Terrax, a wild place only provisionally maintained by the World's Council.

To him the PHDs resembled germs, nothing more than a viral bacteria that should be cleansed from the planet. However, he had been given strict instructions to go directly to his assignment without "diversions", as his boss would have phrased it, and go directly there he would. Always time for fun later.

As he arrived at his exit point, he paused, and with an ever so brief thrill of anticipation, which he instantly quashed, he realized that someone here (possibly all the people here!) may be his target. He felt a warm familiar rush, as he winked in and out of existence between one heartbeat and another, knowing his victim would die while watching him carefully pare the skin off their face, then arrange it in a colorful fashion atop his head.

He would stride out of their pathetic meet-up, his new hologram featuring three heads, each wearing a grisly new chapeau composed of the face of his victims. Hence his nickname, The Milliner.

As he let a tiny smile stretch his flat thin lips, he knew he must punish himself severely for allowing himself that moment of pleasure. The gen-ennges who made him wanted him to be a master torturer, and he was, but he did not know why. And he did not care.

His instructions for the AGGIE meet-up ended abruptly, and he was secretly redirected to a private cubby in the RES105689 sector.

Not changing stride, he headed for the res sector. Making for the entrance of the shabby warren, he ducked his head and flung aside the Targa helmet as soon as he cleared the Environmental Protection Barrier of the building.

The main port unsealed to expose his first hapless victim, as he barreled headfirst into the tiny room, knocking over a short surprised man in a cheap, aquamarine Aerosuit.

Without a word of warning the Milliner reached for the man's throat and crushed his larynx with one powerfully muscled hand.

It gave way easily with a satisfying crunch, and he flung the body aside as a dog would a rat with a broken back. Thinking twice, he grabbed it by the upper arm and carried it into the next alcove of the warren.

He then turned in a slow pirouette, to face the tiny nestling tube in the corner of the cubby. He had an opportunity to pounce again without warning, but waited one triumphant millisecond, just time enough to allow the nestling's caretaker to enter into his field of vision. Then with intentional calm, he kicked it into a million fragments, killing the child instantly.

Now was the Milliner's moment to shine. All these quick, violent actions were only a prelude to this, his best work, the artistic rape and rendering of his final victim.

He would start his torture by forcing himself into his victim's stream of consciousness. The MOB had given him this neat new toy, one that allowed him to jack-in to his victim and record all their responses to his mayhem. This would provide the Milliner with excellent 'gramming later, giving his viewers that compelling moment of shock, as they locked in to the feeling of helpless pain from his victims.

Personally, the Milliner used the tapes to study and perfect his art.

He turned to find her standing stock still, unable to move, her mind frozen with disbelief.

She had time to see Teekum open the air seal to leave and saw the stranger crash through the door. As he broke Teekum's neck and tossed him aside like a piece of trash, she stood paralyzed and watched this creature from her nightmares smash open the nestling's tube, and kill her future.

She knew that she was next, and yet still could not move. Her mind had stopped; she was mute, whimpering and helpless.

The Milliner looked at her and at the ruined body of her lover and had a flash of inspiration. He lightly took the four steps between them, and to her horror, scooped them together, grinding her face into the still warm corpse. She was about to open her mouth to scream and thought better of it.

With a powerful lunge, he pushed them all up against the vidscreen, and begun thrusting his body into them, grinding again and again, mocking the sex act.

As he continued to bruise and pound her, he let Teekum's body slide to the floor in a slushy mess at her feet.

Changing his tactics to give his performance depth and texture, he paused. Gently shushing the mewling mass of humanity trapped between the wall and his unbending body, then rudely poked around to find her bloody neck port with his thick forefinger and jacked-in to her private memcell board.

They were "live".

He could feel the delicious rush of electrons as he invaded her sensory cells, the ultimate in mind rape. She gasped as she was infused with his feelings, every moment of the pleasure he *felt* while he violated her, experienced in brutal clarity.

I see him.

Every time I close my eyes, even for a moment,

I can see him in minute detail.

I cannot shut out the image of the cruel creature.

I remember looking up, and for a moment, clearly seeing his face. He has light ice-blue eyes, and off-white skin. His complexion is unhealthy, a light case of acne on a pale, woeful face.

He has straight, perfectly spaced teeth, almost too perfect.

A grimly smiling mannequin.

His hair is short, a mousy light brown. He has a vacant, blank look in his eyes, as though the ocean of his soul would barely get your feet wet.

He looked at me, yet his eyes never registered my presence.

It was as if, for that moment, I wasn't there. My existence had blinked out.

I believe that if you had asked him a question, he wouldn't

have answered, would've remained mute. Perhaps weeks later he might answer, but it would not be the answer to the question you had asked.

He was in his own world, in sync with no one, at no time. He was beyond evil, he was soulless. When I looked up at him and realized there was no escape, I softly spoke the word "intruder" into our highly sensitive voice-activated security alarm system.

I knew the alarm would summon The Keepers.

Too late to help me. He was coming toward me with a knife right now, and help was too far away.

As he began, I could feel the cold flinty sharpness on my forehead from the edge of the blade. His eyes looked through me as he carefully flensed the skin from my face, and fashioned a hat for himself.

I screamed in agony as he calmly informed me that this is how he earned his nickname "The Milliner".

We both knew that I wouldn't live to retell the story.

Gasping in shock and pain, my mind refuses to fix on the present. I know he has just murdered my heart, my life, my beloved Teekum Dowdy and the nestling. I thought seeing that would kill me right then on the spot, but that would have been a mercy.

He wasn't done yet.

The Milliner ripped through the Aerosuit I had sprayed on that morning, grunting in his effort to hold me with one hand and rip off his own suit with the other. He activated an enormous cybernetic organ that looked more like a weapon than a body part. It was a dead-dry, blunt instrument of pain. It was wider and longer than my outtake orifice, and he tore through flesh and sinew as he shoved it into me. He shoved again and again in dreadful repetition.

In almost unbearable pain from my skinned face, I realized he wouldn't stop lunging into me until my hips were crushed, my pelvis and back broken, if I didn't stop him.

I fainted from the pain. Before I came back to full consciousness, I was vaguely aware that he wouldn't leave me alive, now that I had seen his face.

I fainted again as he dragged me toward an exit portal. I should have fought harder, but he was inhumanly strong.

I couldn't stop him.

Then, he committed the worst sacrilege of all. He opened up an exit portal and, against all societal rules and mores; he tossed me OUT after he was through mutilating my body. He knew that within minutes, the toxic elements would start to work on me.

Radzeeze, the radiation disease, would kill me within days if I wasn't back inside immediately. If not that, then the squamis and scroppers would come around and salvage my still living body for the parts that were still usable for sale.

I closed my eyes for the last time, swiftly remembering the beginning of this day-cycle, how my miniscule problems had seemed so insurmountable.

Now, they seemed homey and comforting.

A million breaths and one short lifetime ago.

Her last memories of this world resurfaced….

* * *

Grist for the Mill

"Lots of people in one big line,
waiting to go through life.
People unwilling to
take a stand.
I've lived in crowded cities,
worked among the masses,
I've dressed as I should
I've bowed and scraped."

- BY THE BARD, PHOENIX GAET

Frankie was in an extremely bad mood this morning, her feet were freezing, she was running low on chedits, and she still hadn't ordered up her usual shot of "Liberty", her favorite wickedly spiked J-juice.

She quickly jumped into the sonic shower and closed her eyes as her Aerosuit was sprayed on every pore of her body. She luxuriated in the convenience of not having to paint on her skinsuit every time she went to her work assignment at the OTEC- Virant Division of the detoxification plant. She carefully chose a hologram resembling an oak tree, down to every detail of bark, skin and sap. It was all the rage down at the plant to "go natural", as they say.

She giggled as she imagined her beloved Teekum Dowdy's face when he crept out of the sleep pod and saw this get-up. It was always such a joy to make him crack that endearingly crooked smile.

Trying to clutch at her wisps of concentration, she forced her mind back to the progress of her department's new thin-film amorphous saline silicon project. They really needed to develop more on-line diagnostic systems in order to monitor the loop control phase and economize the maturation processes of the regenerative fuel cell banks.

Smiling with pride, she remembered how she had gotten to be team leader for the Virant Division. She had pioneered the idea of using old OWCs to pump deep sea pockets of nickel metal hydride concentrates into vast holding tanks, which in turn produced a continuous regenerative fuel cell energy supply the could power the whole plant. Why, the savings in engineering startup and commissioning fees alone had lowered their production costs, and created a huge profit yield for the next three Wheels of the Year!

Things were finally going well for she and Teekum, they had even started up a nestling, whose top of the line "Tiny Tot Tube" was sitting comfortably ensconced in its bracket on the wall, right next to her favorite holopic of Auntie Vee.

As if springing up alive out of her thoughts, Teekum peeped her from his alcove, already clad in his run-of-the-mill holo of the moving sea, which at the moment was flowing over his face in a manner that distracted her. She sighed inwardly wishing he would change that same old holo for a new one, anything else would do.

She also wished he would stop lollygagging; he was already late for work!

Instead, she caught him lazing about, still in his cozy sleepnet, playing with the new robo pet that she bought for him. It was a pint-sized feline model, the CC2O4 – Calico Cat, that they decided to name Tootsie Roll. Tootsie loved to stretch, arch her back, as cats will do while lying down, and then reach up lovingly with her white stockinged front legs, all striped in a tawny gold.

She'd look at you with her quizzical piebald face, one eye surrounded in jet black fur with a streak of tawny up one side of her nose and her other eye rimmed in white with more tawny on her other cheekbone.

Her long back had a saddle of black on it, underscored with more caramel-gold fur. Tootsie would luxuriously stretch her back as you reached out to stroke her fur, and she'd purr in delight, then roll away from you.

She would make you stretch even farther in order to reach her, then gyrating, she would flip back around toward you and roll hither and thither on the floor.

Tootsie would have been happy to do this with Teekum for hours, who had just now arisen and had to get going to his mysterious new work assignment in the Eedia Repair Corp, the experts in the manufacture and repair of planet wide communications equipment.

It was a top secret job, evidently so secret that he had been told to lie about where he now worked, if his previous co-workers asked.

His specialty was tidal mathematics; his previous assignment was to increase the wave power of oceans through the process of "flattening". He truly didn't understand why he had been chosen for this new assignment, but he wouldn't argue his fate.

He desperately needed the chedits. He'd grossly overspent his money on the Senselle Channel, had ended up living in the

Warrens, and was going to be treated as lowly grub so long as he was here. He and Frankie had been Officially Devolved (one of the few couplets who had been "sent down") from the beautiful Melissae Hive.

The Melissae Hive was the most important structure in this "Orb of the Worlds" as the World Council had dubbed Terrax. Terrax, being the center hub of a megalopolis of planets, planetoids, moons and comets, referred to as "the little sisters", was the biggest, most populated and wealthiest in the Galaxy.

Teekum had weaknesses, and even though he was an intelligent Techie, he was also a bit lazy. Except when it came to Frankie.

As he thought of her, a slow sensual smile started to stretch across his lips. He languidly traced the curves of her body in his mind (as he never could touch her in the waking world).

He mused, isn't 100% Virtual Reality a sweet thing?

Viro-Tours Programming could take you on a trip to heaven; lovers experienced each other as delightfully as any couple could have on Olde Earth, without actually having to touch each other!

Physical contact between humanoids on Terrax created a chain reaction in both bodies, causing them to melt into goo. They would be reduced to a mess of leftover living flesh and some electrical wiring, the residue of cyberfolk who are concocted in test tubes.

The outside environment on Terrax was such that if the toxins linked with indigenous living matter, that matter would begin a chain reaction of disimaging, similar to what the TDM (the Thermal Disimaging Machine) could do to a willing PHD.

They had never actually touched each other, for Teekum was born of this world, as was Frankie his meshling. They had had their financial ups and downs, mainly because of Teekum's addiction to the Sextel Network.

He would constantly raid sexual television. (They used to call it pornography, but now due to the fact that it was based on old reelies and feelies from the past, it wasn't considered pornographic.)

After all, it is the interaction of human flesh that can never take place here, but simply watching it is allowable. Nothing is taboo; you can watch anyone, doing anything, to anybody.

The pseudo-humanoid drones, or the PHDs, as they would come to refer to themselves, enjoyed watching.

Humans had become a race of virtual voyeurs. Watching was the only game in town.

Teekum knew that the Senselles had arrived only a few million annuations ago. However, they had immediately risen to the status of Gods and Goddesses, for the simple fact that they had arrived out of nowhere.

Here, on this excruciatingly organized and structured planet, there were no surprises. Nothing was truly mysterious. Until the Senselles came along, that is!

Introduce into this drab existence.....the Senselles, who had arrived and immediately made available live "raw" emotions into the Vidnet via 100% Virtual Reality Programming. Nothing could compare to it – even for a few microseconds- emotions were more riveting than any drug available on the Vidnet.

Any type of entertainment available could not even touch what a Senselle could send you directly over the vids. Even what he and Frankie had experienced in VR rooms before They came, couldn't come close to it.

So Teekum became addicted to the Senselle owned Sextel Network, and soon spent more chedits than he or Frankie could earn in an annuation. Even Frankie, an OTEC hotshot who worked in the desalinization plant couldn't make enough chedits to pull them out of debt.

Sighing inwardly, Frankie ignored the problem, because Teekum was wildly attracted to her, and she to him. She forgave him his addictions and promised him things would be okay again.

She bought him the little robopet as a last extravagance before going Down into the Warrens, in shame and isolation. All of their friends and all of their connections had, of course, been from channels and channels above, but Teekum had faith in their new market share.

He had faith that they could provide superior editing skills for the local Vidnet, and drop the hottest new programming on the vids featuring their new robopet! He wasn't really worried about recouping their losses and evolving back up into the Melissae Hive.

Teekum was so sure of their future this morning, so tipsy with contentment that he had fallen back to sleep on the ruby red divan that he had purchased for Frankie. It was one of the few items he had managed to salvage from their beautiful cubby above and bring down into this dank hole below.

Teekum could hear Frankie yelling at him to "get on the ball", so they could get to work. And knowing he was late, he started scrambling through his mental tricks to try and wake himself up.

But his mind was afloat in a sea of tidal-mathematics, his eyes gazing off into space, as though mentally following his old math Professor Shawking demonstrating a PowerPoint lesson in sixth-space-time density.

(3) " 'The wave power per unit wave front in deep water is given by: J= (p) (gH)^2 T/32 (3.14158) = (a)H^2T,' " (pg 188) Prof. Shawking would emphasize his points by staring out at the monitor as if he could actually see his students.

Chuckling, Teekum brought himself back to the present with an effort. He and Frankie rarely had time together anymore, now that they worked opposite cycles and for different companies.

He wanted to savor this morning, keeping it clear in his mind to enjoy, a bit of mind-candy for later in the day.

Instead, he found himself dashing for the front air seal, to run headlong into the last surprise of his short life.

***

The Ally

"the night was dark and smoky,
he walked thru swirling mists.
he said he was the warrior,
who used no weapons, raised no fists."

- FROM THE ALLY, BY THE BARD PHOENIX GAET

The Retrieval

Cami threw on blue plastiskin pants with a tie dyed T-shirt festooned liberally with dancing Hopitu Shinumu Kachinas. Tuned-in to the local Hopitu Wheel of the Year Network, the dancers gracefully performed the stately Lakon Dance, winding from the front to the back of the shirt. This dance was held in autumn, when it was customary for each tribe member to carry baskets bearing small gifts for each and every member of the community. The dance and subsequent gift distribution was followed by a lavish corn feast to celebrate the last harvest of the year. And then more dancing, far into the night.

Sighing, Cami tore her attention away from her lovely, new animated clothing and turned to the inevitable, though daunting task at hand. She had just received her new work assignment from Auntie Vee, whose memory was as long as the days, and as sharp, sharper in fact, than the tart industrial smell of the canned air in these temporary digs.

Cami wrinkled her nose in distaste at the less than posh surroundings; the clean but dull furnishings of a W.C. issued cubby, with the too small daybed, an outdated percomp and Snack-in-ator, and the eternally blathering vidscreen.

Not for the last time, she virulently missed her old life. For a moment. Then, knowing this was a waste of her mental energy, she calmly returned to reviewing the Retrieval Instructions that had been peeped to her at the top of this day cycle.

Nowadays, she didn't often let her messed up mind dwell on the past. She knew it was strictly forbidden by the Sisterhood of the Senselles (especially by Auntie Vee), she knew it was Best This Way, but at times Cami still missed her life back on Olde Earth. Sometimes she found it almost impossible to push from her mind. She vented another heartfelt sigh. She had been soul searching lately, and that was never a good sign.

Who am I, anyway?

I used to know, but now I can't figure out how to fit into my life here. Of course if you listen to the incessantly Talking Heads, you may know the propaganda better than I do.

I am Cami, Chameleon, the first Senselle, the original. I don't have to fake it, I can Remember. You see, I am a time traveler of sorts (more like a time stumbler), because I didn't get to Now with a time machine or a mechanism of any kind. I was simply in the right place when Time itself ripped apart.

The time stream, which seems one-dimensional to most people, is comprised of billions of vortexes of realities, collapsing and colliding through the space-time continuum. At each twist and turn there exists a multitude of realities, each spinning off in their own direction as prisms through a jewel. As a rule, no two vortexes can collide because of the infinite number of directions they may be traveling in, however when two realities do occupy a single timeline, one of the realities subsume themselves and the stronger vortex replaces the old.

In other words, the Olde Earth physicists who spoke of phantom energy and time ripping apart had been correct, to a point.

When the astrophysicists Parker and Raval identified a fundamental new characteristic of the fabric of the space-time continuum, during the late OO' s on Olde Earth, their rival cosmologists doubted their findings.

Another group of scientists claimed that "dark energy is something we have no clue as to what is causing it, and it doesn't fit into current physics theories, therefore it is a non-issue in the grand scheme of physics."

The nay-sayers couldn't have been more wrong.

It turned out that by researching the ancient master of physics,

Einstein's earliest work could reasonably explain the mysteriously expanding universe.

Although the universe is expanding at an accelerated rate, it is also flat, meaning that parallel light beams never converge. This is possible only if the universe contains a certain amount of "critical matter".

Since our instruments cannot account for all the visible and invisible matter, physicists name the undetected matter "dark matter".

However this still doesn't account for the motion or acceleration of the cold dark matter (CDM).

Phantom energy is what makes up the difference. They call this the quintessence model, which assumes that an energy field pervades space.

They believe this is a new form of dark energy that is providing the pressure to accelerate the universe's expansion. What they don't know is if it is congealing or accumulating in one place, and what its affects would be on the atmosphere around the P-energy mass.

Mutter, an astrophysicist at the Melissae Hive Megalo-Labs, headed up the first group to publish the findings on the accelerating universe. He claimed that the confirmation of P-Energy will create an entirely new brand of physics.

At least, that was the technical explanation, as I understood it from an e-lecture I caught given by one of Wad Medani's MagicoMathematicians.

Yes, we "people" were quite the oddity in this rigidly controlled world.

When I first arrived, I had metamorphosized directly into the UberLaboratories at OTEC (Oceanic Technologies Energy Corporation), where they were undergoing top secret experiments in tidal energy reclamation.

No outsiders were allowed access to the lab, so my sudden appearance surprised everyone, to say the least.

It had especially excited them since I hadn't set off any of their state-of-the-art security systems while waltzing right in.

Of course, Auntie Vee had been assured by The Very Best Techies that it was virtually impossible to pierce the company's security veil.

Ha!

The UpTown boy geniuses, all eager to get their hands on a hot, new project looked at me with the same feral intensity as a cat watching a rodent. They were more than willing to hear the bizarre tale of how I'd unexpectedly arrived in their midst, and when they found out I was from a time when each human had a full range of senses at his or her disposal, they went nearly mad trying to figure out how to capitalize on . . . well. . . me.

After hearing a little bit about their world, and how they were at the mercy of the Media Overlord Barons (the MOB) who controlled all the electronic entertainment media on the planet, I originally devised the concept of the Sisterhood of the Senselles.

Of course, I had plenty of cooperation from both OTEC (Auntie Vee, principal owner) and the MagicoMathematicians (headed by Wad Medani). They were the most famous of all the Senselles, being that they arrived in a pair (in coitus, in fact). Bet that was a shock to the poor old Drones around here!

I immediately proposed that the Senselles develop the rights to our own programming, wherein viewers could jack-in and experience our memories via 100% full sensory Virtual Reality Programming (VRP). The Sisterhood of the Senselles became an instant hit; and the MagicoMathematicians grew rich off their promised percentage of the profits from the Sextel Channel, and for keeping my true identity a secret.

I became the second wealthiest woman on the planet. With the proceeds from the Sisterhood helping to fund Auntie Vee's little pet projects, it was a win-win-win situation, enfolding the OTEC Techies, the MMs and the BayBays into an unbreakable alliance with the Senselles against the all-powerful MOB.

Personally, politics bores me half to death, but I have been told that it cannot be avoided. (Doesn't mean I have to enjoy the cloak-and-dagger routine.)

Luckily, none of this is my problem. My challenge is to continue to bring up memories that I can share with the rest of the Sisterhood, facilitating their unique talents and memories to perform rites of sensuality.

Understand, it is not necessarily a sexual kind of sensuality; in fact the variations on sex can get old quickly, as many of the Senselles who specialize in selling sex will tell you.

Rather than those emotions, I reflect on the ordinary; a mundane grocery trip to the Farmer's Market; or remembering in exquisite detail my harrowing annual dental visit. I can recall strolling in the park on an autumn afternoon or wrestling my dog into the car to go to the vet.

All of these memories require millions of bits of sensory data input; a myriad of feelings, thoughts, tastes, smells and sounds that, to this starved and jaded world, is a rarity worth billions of chedits.

I've always carefully protected the integrity of all my true memories by cautiously doling them out second by second; stingy with whom I share them. This way, I know I can offer the most popular 'gramming around for the rest of my life.

What with the abundance of legal, state-issued, mind-altering drugs available, and too much time on their hands, the poorer DownTowners have to be satisfied with one or two-dimensional programming; reelies, feelies, grammees and cheap holovids.

But anyone living in an UpTown Hive has a higher standard of entertainment to consume. They can afford to buy 100% VRP, expensive skinsuits that look like real clothing instead of cheap plasfoam sprayed on your skin, and even purchase high-grade protshots suspended in pure water.

These, Cami swelled with fake pride, These are the people who are my fan base. My beloved BayBays can jack directly into my brain and my memories, if they can afford it. And as far as I'm concerned they're welcome to them.

* * *

Coming back to the present, forcing herself to mental clarity as she had been taught by Esthana Thlehiya, Cami allowed herself only one moment to feel her distinct distaste for retrievals.

They were dangerous, chancy, high investment/low return assignments that usually ended up making her feel melancholy, which made her feel nervous, which made her careless, which usually ended up with someone getting wasted.

And in this connotation, getting wasted WAS NOT a good thing.

Esthana Thlehiya, her Hopitu Shinumu Shaman, had taught her to take the time to revisit a strongly imprinted memory; by doing so it would focus her mind, slow her breath and help her to begin a self-induced hypnotic process. By the end of the process Cami would again be in complete control of her mind, body and spirit, allowing her to perform her upcoming retrieval in peak alpha state.

Taking three slow deep breaths, filling her belly with clean air, then exhaling to a count of ten, her eyes dulled as she turned her mind inward, reaching back to a time when all her thoughts belonged to her own secret, private self.

Cami felt the familiar tingling of energy as her chi began to spiral up her spinal column, slowly filling each chakra with the appropriate color and vibration of energy, before flowing up, up, up her spinal column, until finally peaking through the crown of her head.

She shut her eyelids, hard, once. It was her signal to her subconscious mind to *Remember.*

* * *

All at once, she was driving her drop-dead gorgeous red Grand Am along, heading eastbound on the 210 freeway. Cruising along, singing softly to her new Clannad CD, she felt happy and secure, confident and in control. Her beloved San Gabriel Mountains on her left, standing at a medium height of about 6000 ft., were colored a dun brown shade, tinged with an occasional stand of yellowing liquid amber trees. These mountains were filled with ancient oaks, massive and silent, and pine trees of all shapes and sizes. The mountains had always comforted her, like the older brother she never had, standing over her to protect her.

As she whisked neatly down the off ramp, indicating her left turn onto the surface street, an idiot in a thunderously loud Hummer14.5 cut her off, nearly shaving a few coats of paint off her sporty little racer. She cussed venomously, cursing at men drivers in general and this one in particular. After encounters like this, she momentarily wished she could permanently escape this useless madness that she lived in, and thought it couldn't happen soon enough.

* * *

Her concentration broke.

"If only I knew what was coming," she lamented, coming back to the present. "If only I had woken up in time to appreciate the beauty that was in the world."

Truth is great, it just comes a bit too late to be of any use, Cami concluded.

She allowed herself one more brief visual memory of the drive she had taken so long ago, in another world, another time, before shutting down her 'gramming output. She unhooked from the grid.

Stealing one more precious millisecond in private, she remembered the way the sunlight had slanted through the trees that early morning. The light was the color of the oaks in autumn, turning it a rich, warm yellow/green, dappled with shadows.

She inhaled a hearty breath, and felt the warmth of the sun on her skin. But duty called her back, and she had had her moment in the sun.

In this reality, one minute in the sun would melt the skin off your bones, and then melt them too. She grimaced, enough of this sweet self-torture, she had a job to do that required her full attention. She was ready to go.

Cami took a succession of quick, shallow breaths to gather energy, focusing her mind and body for the dangerous bit of flying she was about to try. She loaded up the 'gramming her moplane would need via remote, entering the lat's and lon's for the retrieval pick-up site, calculating an error margin of only 20 milliseconds.

After that, the human creature she had been sent to retrieve would melt into a puddle of icky goo; dead and completely worthless to anyone.

She found herself uncharacteristically scattered this morning, trying to run her hands through her hair, a mannerism that made her feel a little more relaxed.

Clanking her gloved hand on her flight helmet, she had forgotten that she had already put it on while going through free fall to the connector lines, which would take her to her assigned launch exit portal. There her moplane waited, primed for take-off.

Boarding her 'mop, she tried to play the Hopitu Shinumu warrior's game she always played when leaving for a dangerous mission. Tensing all her muscles, she prepared her body to be cat-ready to leap into her pilot's seat at the first possible nanosecond, not quite waiting for the mop door to slide open all the way.

The game had been devised by Esthana Thlehiya, and was used to prepare the perfect warrior to jump into any situation with a 360° awareness, mind awake and unafraid, body capable of instantly reacting in whatever manner necessary.

Rarely, as on this occasion, did it happen that she was not completely ready.

Her wandering thoughts blew her concentration, causing her to hit her right shoulder on the leading edge of the door as it slid inward. Running at full charge, she twisted, flipped around three times, then balanced on one foot and shot into her comfortable pilot's chair. The safety net snaked around her with python strength, then eased back a bit, adjusting to the g-forces of the moplane as it sped silently to its dark destination.

Rubbing her shoulder somewhat ineffectively through her Aerosuit and special enviro-armor, she made herself ready to dash out with somewhat more grace and hopefully less pain.

The 'mop landed right on target, and right on schedule. She trotted lightly on her feet until she located the correct connector line. She jumped into it head first, which was not recommended, (therefore the need for the helmet) and arched her body as Wad Medani had taught her for maximum control of the connector line.

She considered connector lines her "ally", somewhat in the ancient fashion of Carlos Castaneda's writings. Knowing that it was dangerous, she had demanded to be trained in the art of using the lines of energy as her personal transport.

She recalled the lines of her favorite poem, "The Ally".

The Ally

The night was dark and smoky,
he walked thru swirling mists.
He said he was the warrior,
who used no weapons,
raised no fists.

The Indian man wearing a hat,
Shuffles along in old huaraches,
the leather ragged, straps hung loose.
He looked so worn with cares,
I thought his battles were with ghosts.

I doubted him, the deep brown eyes,
his sometimes smile, his riddles of the ally.
One day in hot sunshine, hat pulled low,
we walked on dead river's path
to meet his enemy in time.
The wills were strained between the two,
they fought the other's weakness.
As battle raged into the night,
the veil began to thin.

This enemy is ancient,
and frightening to behold,

it is the oldest sin.

It is the darkness from within.

-BY THE BARD, PHOENIX GAET

Cami rolled the poem over and over in her head like a disjointed mantra, using it to calm and center her. She knew that as soon as she exited the portal, she would have to hustle, if this mission wasn't to fail utterly.

As she landed on the ground, she rolled, unfolding into a running position, then slowed down to get her bearings. Cami crept cautiously along a busted up street in a back alley of Suiterville, so named for the old Aerosuit Technologies factory that used to take up hundreds of square kilometers here AboveGround.

The disturbed, twisted buildings hovered above her, shutting out the wan W.C. issued LED lights. Menacing scatterstorms shuddered in the air around her, threatening to come shrieking down out of the sky at any moment, bringing a particularly gruesome ending.

With an enormous smile that split her face from ear to ear, she was getting that jolt of excitement she always felt when she was attempting to cheat death.

She was Where She Wasn't Supposed To Be, the thrill of fear singing along her nerves made her feel like one of the last real pioneers. Risking exposure to the terrible environment that humans had brought down around their ears was something even the bravest among the PHDs did not dare to do.

She would think of those times when she came from, and knew with certainty that absolutely no one indigenous to this planet would believe her if she told them what it had really been like. Her personal story was convoluted, mixed up and multi-layered, the realities having shifted so many times, it confused even those who had lived through it.

But that was another story for another day.

Right now, she must concentrate on her search. She knew she was getting closer, and running out of time. Scanning her holo-reads from the inner rim of her safety visor, she turned left. A few paces later, she stumbled upon the smoking, stinking blob of flesh that she knew must be her target.

Even with prior knowledge of the disposition of the target, it still came as a shock to see it laying there, exposed to the toxic elements with absolutely no plastiskin or protection whatsoever, for who knew how long.

Cami immediately sensed something was wrong with this retrieval, and wondered if Auntie Vee hadn't made an error.

This...thing…didn't seem like it could be salvageable. Disgust instantly turned to resolve, and Cami was determined to give this her best effort.

For some reason, she felt a sudden, striking connection to this being, whatever it had been. She tried to look at its face, and for a horrible second, thought that it had truly been OUT for too long.

The face was completely unrecognizable as a human face. As Cami continued to stare, to her fright and horror, she realized that in fact, the person's face had been skinned off.

"What in the world happened here?" she asked herself, aghast at what she saw.

On impulse, she grabbed up the stinking mess, and decided to go see Auntie Vee about this, personally.

She had seen a lot of things happen, working as a Senselle, but she couldn't…no, she wouldn't walk by this. She didn't know why in the world she decided to get involved, but she had made her choice.

Cami, being who she was, had the power and money to be able to make that choice. Only one in 25 billion people on Terrax could have been lucky enough to be retrieved by Cami. The chances were almost nil.

One thing was for certain. If this person lived, whoever it had been, that life was over. This single act had changed it all forever, now they were destined for another's purpose.

Fate (in the shadowy forms of Auntie Vee and Wad Medani) had had a hand in it. Of that, Cami had no doubt.

But, why? Why?

Shaking her head, half dragging, half carrying the remains she had placed in a black body bag, she got back to her mop', loaded up her eerie cargo and sped off to Auntie Vee's complex.

Cami hit the latch-down seals and the mop' ripped off its moorings with a vicious force. She had only seconds to arrange her body as the automatic safety netting settled down around her and locked her in. She made sure that she had a weapon activated in one hand and the other hand on a manual control lever, allowing her to take over control of the moplane as she landed. If necessary.

Blazing through the coriolis winds, she navigated around the deadly scattershowers, composed of toxic metals formed into whip like threads kilometers long, capable of tearing through the moplane, had it not been exuding a disteldt force-shield.

Cami knew that only a few of the Senselles were authorized to have the DF shield on their personal moplanes, which was some little comfort as she simultaneously kept contact with OTEC's comp tower instructions for landing clearance, while monitoring their constant chatter with one part of her brain. All the while thinking about what she would ask Auntie Vee once she got there.

With insanely fine-tuned reactions, she jetted around and through various obstacles, as her GPS sector chief reported the worst of the weather conditions, and cleared her arrival at OTEC in a matter of nanoseconds.

There was never time for the important stuff!

Chagrined at herself, she also had to admit there was another good reason she was having trouble concentrating. She had recently jacked-in to the Senselles Home Channel, having been allowed the rare privilege of accessing one tiny crystalline moment of real time from Auntie Vee's relationship with Wad Medani.

Cami had since become mentally derailed, knowing that she shouldn't have messed around with this particular type of VR programming. As much as Auntie Vee had lectured the Sisterhood on avoiding that topic, Cami found herself drawn to the forbidden fruit.

She found herself obsessing over the whole concept of "romantic love".

* * *

As Cami loaded the body bag onto a conveyance vehicle, her mind went over the familiar argument.

Yeah, I know I've felt love before, but I've never experienced the bitter- sweet complexity that Auntie Vee feels for Wad Medani. It's hard for me to wrap my mind around the idea, because the emotions are all mixed up; she feels a warm compassion mixed with intense rushes of anger; a familiar love interspersed with primal sexual pleasure; a sense of loss, of deep loneliness when not in his physical presence, all combined with an innocent, maternally-fierce possessiveness of him.

I've never had these feelings for any man. In its place, I've had the unfortunate experiences of having been beaten, raped and mentally abused by the men in my life. I had become extremely bitter about the whole subject of love in general when the Time Rips forced me into this future world.

I wasn't looking for love or lust, and Now, here, there is nothing sexual at all in my nature. I won't allow it.

Besides, I'm here to do Auntie Vee's bidding, having only a slight fascination for things here (like the AGGIEs). Unlike the world I came from, here I'm important. Here I can get anything that I desire (at least anything that this planet can offer).

Now, I've gone and made myself unhappy, because now I know the difference between what I thought was love and what love really is.

Now, I wish that just for a moment, I could feel love the same way as Auntie Vee and Wad Medani.

One thing I am curious about is love felt the same from Wad Medani's point of view. Does he love Auntie Vee this same way? Or is it different for men?

Unfortunately, that's the one viewpoint that I've been denied access to, because Wad Medani steadfastly refuses to have anything to do with the Sextel Channel. He had originally agreed to sell his technical knowledge as programming, but he'd not allow his personal memories to be jacked into the Vidnet.

Wad Medani's probably the only human on Terrax who has that privilege. Although he was able to refuse the public access to his personal feelings and memories, they must still be recorded directly into the TDM. That was in the original contract that he signed with Grandpa and the MOB, in the long ago, when the Senselles first arrived.

When we first got here, this planet was rigged to work only one way. The W.C. supplied the viewership with a cubby, protshots, clean oxygen, and basic interactive Eedia, but not unless you were hooked into the TDM. That's the way it worked.

Only the most privileged and powerful have access to the direct output of the TDM. As it happened, the MOB was the most powerful group in the land. That is, before the Senselles had arrived.

Once the Senselles came onto the scene, Auntie Vee, Wad Medani and his gang of MagicoMathematicians, (all of the techie people who had shown up from the past), became the new power group.

Of course, I was glad to be welcomed into the ranks of the few and powerful. There were so many emotions that I'd never experienced before that I have access to now, like this most recently tasted emotion "love".

It was too distracting, too overwhelming, too all-consuming.

(And she couldn't wait until she could afford just one more nano of Auntie Vee's memories of it.)

In the meantime, her DP exit had deposited her at the doorseal to Auntie Vee's Uberlabs.

* * *

Cami sighed as she thought about love and life, then wondered again about what was left of the poor creature in the body bag. The body bag, although rather crudely named, carried within it intricate mechanisms to keep the being inside it alive, if only for a short time.

No time left for thought, Cami soon found herself at the top of OTEC headquarters, the largest megalocorp in the world. It was Auntie Vee's private lab.

The connector tube whisked her up the top levels of OTEC HQ where a crew of MMs were ready to take the body (or what was left of it).

It would be their job to see if they could revive it into living flesh (that was more than slightly tainted with enviro-toxins.)

As soon as the auto doors slid inward, Cami realized the gig was over, that she would have absolutely no input into these taut, precise goings-on around her. She was given a curt nod by Auntie Vee, who signaled her over to a couple of OTEC-Panatechies, who were preparing a programming jack-in for her.

Cami, with bitter understanding of the double edged sword of success, must give up yet another fragment of her soul, of her past, of her memories.

Her fans were beckoning her.

Now.

As a rule, Cami didn't like to think about her life as a teenager, it was miserable and filled with violence and domination. But today she craved those carefree days, days when she could hitchhike down the street, in full sunshine, with not a thought of danger in her head.

As long as I have lost myself so completely, why not remember it all, right back to the beginning, when I was truly free? Let the viewership suffer. Let them know the beauty of being young, crazy, fearless! They'll never forget it.

And they'll never get to experience it again.

She derived some small pleasure from the fact that she wasn't doing them a favor by "sharing" this memory. Not this time.

She began to *remember.*

* * *

MARDI GRAS- Done the Hard Way, Part 1

One Thursday afternoon, during the Summer of 71', Cherry came over to my house without calling first. This was unusual because she lived over a quarter mile from my house and always called before coming over.

Cherry sauntered up in her no-hurry way, smoking a cigarette and wearing a tiny frown. She doesn't talk a lot, she keeps both her frustrations and fears to herself, most of the time.

She's much taller than I am, with long, straight, dark brown hair and thick expressive eyebrows. Her light complexion sets off her brooding brown eyes nicely, but today the effect was marred by the dark circles under her eyes.

She had that angry look that told me her step-father had hit her with his cane, again. That, and the slight wince she tried to hide when she put her weight down on her right foot. She looked grim.

Her step-father was a raging alcoholic, a mean spirited, small-minded man who liberally used his walking cane on his three step children. He thought life had handed him a bad break.

It was getting to be a daily thing now, and although her older brother tried to draw off her step-father's rage to keep the younger two from getting the beatings, it was to no avail.

I could tell things had reached critical mass.

As she leaned against the garage door, eying me without saying a word, she saw the same pain mirrored in my face.

Standing at 5' even and weighing in at only 101 lbs., I could scarcely be considered threatening.

Yet I had once again injured my father's delicate ego and overweening pride, and had paid the price.

As she took in the various bruises on my arms, and the anger in my flinty stare, she knew that I too had "hit the wall" a few too many times.

My hand unconsciously rubbed my jaw where I had been whacked for speaking "in support of the gay community".

I smiled wryly at her. She knew exactly what I was thinking.

We had talked about splitting for weeks. We knew we would either have to kill someone and end up going to jail; or, we had to get out of here.

We made our plans to leave.

But, where could two under-aged girls, only 15 and penniless, go? We had no jobs, no money, no way to make a living, no one to help us and no way to get away, even if we had a place to go. So, what was the use?

Then I remembered.

I had been writing to a college guy named Ben, who I met last summer. In passing, I had mentioned my problems with my father, and teasingly asked him "What would you do if two vagabonds showed up at your door?"

In his last letter, Ben had replied "You can come to see me anytime!"

A hopeful plan sprung to mind.

We figured we'd pack a change of clothes, steal any cash laying around, and hitchhike to northern New Mexico.

We were eventually headed to Mardi Gras for the party of a lifetime, but figured we could first visit Ben, regroup, and then find the fastest way to New Orleans.

THE MIDNIGHT FOOTBALL LEAGUE

We had badly underestimated how long it would take to get there. Around midnight of the first day we were tired, hungry and nearly frozen. Our last ride had dropped us off in the middle of the night, and here we were alone in the California desert, somewhere around Needles, without coats or even a sweater.

(It would have looked a little suspicious to my parents if I had left the house with a coat at 4 o'clock on a sunny afternoon. So I had grabbed my purse with the few dollars I'd saved, my address book and a map, and split.)

We tried to keep moving, but were stumbling wearily along and had to stop and rest for a bit. The night was getting progressively colder and Cherry was shivering and turning bluish around the lips.

We were perched on a huge rock embedded in the sand at the edge of the highway. I didn't want to stray too far from the street light, into the forbiddingly dark desert.

I started to panic a little, and spoke louder than I had meant to.

"Cherry, come on. Don't keep sitting there, you're turning blue!" I urged her to stand up. She just stared at me.

"Hey, let's pretend we're playing football. Come on, go out for a pass," I drew my arm back like I was going to pass her the ball.

"I can't m-m-m-m-of," she was shivering uncontrollably.

"Get up! Hey, here comes a car, maybe they'll stop!" I was pulling on her arm more insistently this time.

She grumbled, got up on her feet and gave me a dirty look. Just in time, we realized that we did not want a ride from these particular guys, and ran across the highway to the other side of the road.

Several drunken men were hanging out of an old, beat up lookin pick-up truck, yelling loudly that they were gonna "rape us good". Suddenly, their truck made a lumbering U-turn, they were coming back for us!

Hearing a noise behind us, we spun around to see a couple of young guys getting out of a camper we hadn't noticed before. It was parked on the side of the road.

"Are you girls OK?" one guy asked, concern evident on his face.

"Yeah, I guess", this was from Cherry who had good reason to distrust men.

"No, we are not OK! Those guys are coming back for us right now!" I was too frazzled to make much sense. "Hide us!"

I instantly trusted these new guys, they entirely lacked the evil looks of the men in the pick-up. I had no foreboding about getting into their camper.

The guy who had spoken first was tall and spare, and spoke sincerely. And, I was freezing, so his camper thingy was looking better and better.

"Go on, climb in the back, we'll figure out something. That is...if you want to?" his blond friend offered hopefully. Cherry looked skeptical, but she was always a sucker for blonds.

We stumbled into the back of that lovely, warm camper shell as the guys took off into the desert like they knew what they were doing. Cherry started looking a bit more animated as she warmed up.

After awhile, Bruce and Don (our rescuers), stopped the truck and climbed into the camper shell with us. They told us that we were parked next to the unofficial town dump, where people left old furniture and stuff.

They seemed to think we were safe and wouldn't be disturbed at this time of night.

They asked us to stay with them for the night (what was left of it). I told them we weren't taking our clothes off, but we'd love to cuddle. I guess that was good enough.

For one thing, by now it must have been two in the morning and we were exhausted from the frigid cold, not to mention the exciting chase through the dump.

The last thing I heard before dropping into a dead sleep was the braying of a wild donkey.

* * *

The jack-in ended.

The room started to come back into focus, and Cami saw that she had 'grammed a classic. The return quotients (RQs) on the vidscreens were going nuts. She was a star, again.

And it sickened her. Again.

She looked at the quiet proceedings around her and knew her presence was no longer needed. Leaving quietly for her private quarters within the depths of the Melissae Hive, she felt weary and eternally sad.

* * *

Auntie Vee's Crystal Heart

Once I was a child, laughing at the world.
I knew that I would learn and love,
to live and give, and- win!

I boldly walked, then danced, then ran,
so free and strong and brave.
I knew that life was full of wonder,
and I was not afraid.

-BY KAYLA RACONTEUR

As Cami entered her lab, Auntie Vee carefully studied the face of her favorite charge. Seeing concern, and a kind of pushy determination there, she schooled her features into a strong assured mask.

Auntie Vee was waiting calmly, her usual entourage replaced by a curious gathering of folk, indeed. Wad Medani, whose presence was felt only by his huge image superimposed on the vidscreens to her left, had his minions present in plenty: Bepko da Silva, his number one shark; and the infamous MM ReGen Specialists, Severn and Fundy.

In addition, three of Auntie Vee's favorite servitors from OTEC, Olulpana, Nuupapana, and Naxopana, were scuttling around ID'ing everyone with retino-MRI scanners. To Cami's dismay, her sister-Senselle Damaria stood in a corner "live recording" the stressfully quiet proceedings. Damaria was assimilating every detail, right down to each participant's individual autonomic responses, thoughts, reactions and real time memories.

Damaria was a Senselle with one of the highest rated artiste 'grammes going, performing as a Temple Dancer of The Odissi Hive. She was also a valued employee of Auntie Vee, someone who could be trusted to get a job done, and keep her mouth shut about what she had seen.

That in itself was significant, but Cami was too overwhelmed by all the people present to be able to process the information.

It was almost as if the scene was meant to induce the ultimate confusion of senses and thoughts.

Cami had a sick feeling that by the time she figured out what this was all about; it would be too late to stop it.

Cami had no time for questions, as she curtly nodded once at Auntie Vee, briefly acknowledged the vidscreen image of Wad Medani and finally, to her sister Damaria. She watched her control of the situation recede from view, as the bag containing what was left of a human being was emptied onto what looked like a mini-replica of the TDM.

Desiring more of an explanation, Cami glanced back at Auntie Vee, who was momentarily caught in a distant memory of her own.

* * *

Auntie Vee gazed longingly up at the vidscreen image of her beloved Wad Medani, reminding herself exactly why she was allowing this to be done to another human being.

Auntie Vee and Wad Medani were deeply in love, yet could never again be in physical contact as long as they lived on this forlorn planet. After the incident that had catapulted them both into this time line, they discovered that when they had physical contact, they would age at an unnaturally accelerated rate.

They discovered this quite by accident early on, when Wad Medani had gone to the other side of the planet to fix a serious engineering breach in the Southern OceanDome.

The job took a few years, and they had kept in contact by Vidnet and virtual reality jack-ins. By the end of the project they realized, with a start, that they had not aged a single day since they had been apart.

It also became painfully apparent to them that anyone living Planetside would eventually be infected by Radzeeze, the lethal radiation disease that penetrated all known substances. She and Wad Medani then made a seemingly simple decision that would not only affect their lives, but the lives of every living soul in this galaxy. You see, once they permanently separated and took up residence in their respective space stations floating high above Terrax, they truly became as Gods.

Living high above human kind, never aging, yet old and wise, seeing all and knowing much, they had become legend. It was a bitter choice for them, because although they had almost unlimited power over the viewership, the politics and economics of this world, they had lost each other.

Most times she deflected her mind away from the old memories, but today she felt strong enough to reminisce.

* * *

Long ago, it seems almost eons, we met on Olde Earth. A mutual friend of mine, Melanie, introduced me to an eclectic world-beat rock band called Ka-POW!

The leader of the band, Wad Medani, was sitting around a weather-beaten, yellowed Formica kitchen table with the rest of the band when I walked in, sassily flipping my waist-long jet-black hair over one shoulder. I walked right up to the chair next to Wad, sat down and looked straight into his deep green-blue eyes.

He peered up at me through white-blond bangs tipped in black at the forelock, producing a startlingly vivid affect.

I stared intently at him for another brief moment, then plowed forward, unfazed.

"Wanna smoke a bowl?" I asked.

He broke out in a giant smile that made me feel like the sun had just come out from behind dark clouds. "Sure, sweet thing. Annnnnnd, exactly who are you?" he ventured. I seemed to have all his attention.

"I'm the girl your mama warned you about," I replied absolutely deadpan.

At first, I caught him off-guard, and the nonplussed look on his face made me break character. As I started laughing, the whole tableful of stoned rockers disintegrated into giggles, guffaws and a well-meaning snort or two.

My friend, Melanie, just about busted a gut laughing, and

then we all proceeded to get hysterically wasted together.

And that was how it all began.

* * *

Auntie Vee came back to the present with an infinitesimal start, realizing she had let her attention lapse at a critical juncture. But she couldn't help herself.

She vented a quiet sigh, thinking of those good old days when no one had to worry about getting a disease simply from casual exposure to the environment. She remembered when people could touch each other without transmitting a deadly disease, when we all had the ability to feel the entire range and complexity of human sensations and emotions.

She wondered if Wad Medani was thinking along the same lines. Lately, he had been so preoccupied with the water shortage, the enviro-leakage from the Ocean Dome into the warrens, and the World Council intrigues, she wondered if he could possibly have any energy left over for their little experiment.

He often let their personal relationship go to the bottom of his endless "To Do" list, but she wondered if that was expediency, or had he really given up on Them.

Had he become comfortable being on his own? Did he want them to succeed here...or fail? She wished he'd give her a verbal cue or subtle hint, but as she scrutinized his face, she realized she couldn't read his mood at all.

It pained her to admit that to herself, but it was true.

We are becoming strangers to each other. This morning in particular, I feel an overwhelming loneliness for Wad Medani. He is mentally far, far away from me now and it strikes me with sadness at odd times.

Dammit! Neither of us can afford to be distracted at a time like this.

I'm not one to make the mistake of thinking that merely being busy is the same thing as having actually accomplished something. It is simply not true.

Speaking of which, my e-genda is screaming for attention,

as if I don't realize I have a million and one things to do today. I'm hooked into the Senselle Channel and the Sextel Network, of course, monitoring my various beloved Sisterhoods.

This morning The Cook is on, making a delicious gazpacho.

Which brings to mind our most urgent planet wide problem, our lack of water! That is precisely why the Sisterhood's plan to influence the W.C. must work.

Tonight Damaria will dance, for all public intents and purposes, to perform a "sexual healing" for Laird and Randi; of course, I have other plans for them. Laird must be brought around to see our point of view, and agree to vote in favor of an independent H.O.R.S.E. expedition.

I don't think any of the viewership really understands how desperate the situation is over water. Wad Medani and I know, being among the few people who survived the World Water Wars and came through the time rips. We know exactly what it is – what it does to people- to be without water.

All of our Recycling Processors, who are overseen by Wad and his MMs, are working overtime to refurbish the sanitation machinery; industriously developing new nanobacteria that will clean the filtration systems.

But the "water" here has been recycled so many times, it's almost as though it's another substance at this point. So Wad and I are plotting a way to get our own oxygen recovery expedition funded through the World's Council.

Hitherto, the MOB has pretty much monopolized the services of the Brotherhood of the H.O.R.S.E. (Hydrogen Oxygen Recovery Specialists, Elite). Long ago, the MOB nailed down the patents, the rights, and then the means of retrieving virtually all raw materials found in galactic W.C. space.

It turned out to be a prime power move. The MOB built a huge space station out beyond Phaeron; it was one of the first factories in space to reclaim solid materials. There, all raw materials are screened for solid metals and (especially precious) the grey water that is turned into fuel, oxygen and potable water.

We have been forced to buy water, liquid hydrogen and raw materials from them ever since we arrived. It's not that I mind the cost, although the MOB has been punishing us with multiple rate increases these last two annuations, it's simply a matter that the water demand of 26 billion people has outstripped even what the MOB can supply.

And, although we've gen-ennged the PHDs to require a minimal level of water for their memcell board, we've still projected shortages that will hit catastrophic levels within the next few Wheels of the Year. We simply can't afford to let the water supply fall below a certain level.

Auntie Vee rubbed her face wearily, trying to relieve the stress headache that was starting to pound at her temples. She had to think this through!

The trouble is, Laird (as titular head of the W.C., he is the final word on the votes) and Randi, his meshling, have been at odds lately.

Laird originally fell in love with Randi, fully aware that she is a Senselle who specializes in sex. He was well aware of her wild personality and her unruliness, yet he still allowed himself to become addicted to her.

Unfortunately, now he doesn't want her to continue her work, and there's absolutely no way she will stop 'gramming at this point. We've tried to explain to Laird that it's not interactive sex. We tell him she's only accessing existing memories of her previous sex-capades, when she worked as a masseuse in a swanky Newport Beach Spa.

At 5'9", the rangy redhead drew raves from all of her clients. Her hair is long, about waist length, straight, a reddish brown auburn – really more brown than red. It drapes and shimmers in straight lines down to the small of her back.

Her eyes are light brown, her eyelashes dark and bushy. Her eyebrows draw a sophisticated line that accentuates her flawless cheekbones, drawing attention to the fine structure of her face. Her complexion is light, with a pink glow that redheads can have, although without the freckles. She is slender at the waist, and extremely well endowed.

She stopped making public appearances, once she became Laird's meshling. She's a rather private person, aside from the obviously public persona as a sex industry worker.

It turns out that from the age of nine she had been sexually abused by her father and an uncle, so she decided she would never trust men. However, she did love to use men.

She intuitively understood that her sexuality her sensuous body, her long legs, her full lush breasts, enticed men. And she desired nothing less than complete power over them. To ultimately conquer and then crush them, to see the desire come into their eyes, only to steel her mind against them.

She could make herself simper and vamp, knowing it would turn them into putty in her hands. This, in turn, disgusted her and she would unmercifully use these men for their money, prestige, power and costly gifts.

She could perform sexual acts that even they didn't know were possible, and bring them to their doom, turning them insane with lust and possessiveness.

She smiled coyly, storing this moment in her mind, as she was mentally smashing the man under her heel. She would suck and sigh, internally tamping down her acts of self-degradation, crushing them deep within her, with perfect self-loathing.

When she came through the Time Rips and the Sisterhood found her, she was a perfect candidate for the interactive sex channels. She had had every type of sexual experience imaginable and had no problem sharing them in vivid detail with the viewership. Since most PHDs couldn't afford an entire session with her, she would always, and in perpetuity, have her lurid stories to tell. Nano by naughty nano.

Vee took a mental U-turn, reconsidering the quandary with Randi and Laird. Once the full arsenal of Randi's sexual energy had been unleashed on Laird, not only did he love it, but he didn't want to share her with anyone.

This was okay with Randi, because after all the multiple partners she had had, being monogamous seemed kind of kinky.

At least, it was okay with her for the first few annuations. Now that Randi and Laird had been monogamous for over a decade, she had become excruciatingly bored with him. Unbelievably, ever since they had been together, Laird had never jacked-in to her 'gramming!

Once Randi was bored with their sex life, that's when Laird started surreptitiously jacking-in to her 'gramming to get some ideas on how to spice it up in bed. Bad timing on his part.

It seemed he didn't really know her at all, from what he was seeing. Hers was the soul of the untamed, the outrageous, the completely immoral.

Or rather, amoral, because Randi simply didn't think one way or another about sex acts of any kind. She was ambivalent and unruffled when he angrily confronted her about her last 'gramme, "Sands and California Dreamin".

She doesn't care what he thinks, and that's when he became insanely jealous. That's when he insisted she stop her 'gramming.

And that's when I was called into the picture. Randi usually doesn't come to the Sisterhood for much of anything. She's such a private person, she didn't want to share this problem with the other Senselles, but came directly to me with it.

She came to me because they simply weren't getting along and she was worried about her upcoming assignment to "convince" Laird to get the WC to fund our hydrogen oxygen recovery project.

She thought it would be difficult, to say the least, to rely on her usual powers of persuasion, if they weren't having sexual relations.

Randi agreed with my suggestion that Damaria Rotiarte (The Odissi Temple dancer) invoke a sexual healing for the couple. They would both have to be in Damaria's presence as she danced around them, gathering the potent Ka' healing energy.

Laird had hesitantly agreed to it, but only if nobody would be allowed to see the couple's private responses to the dance. That was the deal Laird cut with the TDM Admin.

We told him we would transmit only Damaria's dance, their private responses would not be made accessible to the general viewership.

Thus far all the work Damaria has done for the Sisterhood has been of such high caliber, that I trust she will do an excellent job convincing Laird he cannot be without Randi, and that Randi will only be available to him after a certain successful vote from the tables of the WC.

(Part of the dance is an autosuggestion to Laird that he swing the World's Council votes and influence a positive outcome for our expedition). This all sounds very complicated, but it's the way things are done around here.

I know, I know, originally I was talking about being lonely for Wad Medani. And I end up talking about our work, but our work brings us together in the only way that we can be together, right now.

One of Wad's MMs, Bepko Da Silva, has devised a special program for us, wherein we jack-in to each other via the Vidnet, while recalling the same memory at the same exact moment.

In order to simultaneously stimulate our memcell boards, we have rigorously practiced an exacting form of mental control, enabling Wad and I to start a memory at the same exact nano.

When done correctly, we are able to sync up perfectly, nanosecond by nanosecond, exactly as we lived the memory. Transported back to that time, we feel as though we are living the experience over again; whether it was going out to eat Greek food, relaxing together in the hot tub, playing around with our dog or leisurely making love on a Saturday afternoon.

Before each session we talk to each other about the experience, remembering each data of sensory input in vivid detail. We discovered a fascinating characteristic about memories.

Apparently, there are details that he'd remember, that I didn't, and vice versa.

Through intense, detailed discussions, we found that our individual memories recorded the scene in its own way. Each of us remembered a different experience; each had felt different sensations while living out the same moment in time.

By talking out the scene we plan to relive, by almost scripting it out and remembering it together, we fill in the picture so that it's complete. Once all the sensory input is there, it's truly a three-dimensional experience.

Even with this highly sophisticated computer program, I still miss the real Wad Medani; to be in his presence, the way he'd smile suddenly, then come over and hug me, kissing me gently. I miss his touch and await with an anxious mind, watching Frankie's progress to see if she will be capable of fully regenerating into a sentient human being.

If the experiment works on her, then Wad Medani and I will try it.

Goddess help us!

We don't know if we'll survive the experiment, or what will happen to Terrax in the ensuing aftermath.

We just know that we have to try.

Wad Medani's Tale

Wad Medani was catnapping, dreaming of her. It brought to mind the bit of poetry she had written him, about a lifetime or two ago.

deity

Tell me of your deepest fears, your
vivid dreams and fantasies.
Can you tell me of these?
Must I force you to reveal yourself,
to give up that which you hold dear?
Can't you see I will know anyway,
without you whispering a word?
My role is not to answer your every
question with a yes,
nor to fulfill your every wish.

-By Vee

Wad awoke with a start. He found himself in his 'mop, on autopilot, descending gently toward the surface of the Southern Terrax mopad, and his meeting with Asna J.

It was too damn early in the day-cycle. Wad Medani had woken abruptly from the wisp of an enticing dream.

It was if some fae creature had lured him from his comfortable bed, and caused him to wander about all night. He felt exhausted, without understanding why.

Too early, and not enough j-juice in my bloodstream, Wad answered his own question. He grimaced as he considered the situation.

I finally figured out the problem I've been having with Vee.

Although I do have emotions, I feel them inside me. Since I don't show them to the world, (meaning Vee) she thinks that I don't feel them.

Faugh!

She and I have gone through this intellectual discussion over and over again, but it's almost impossible to convince the woman that I love her, unless I am physically with her.

(And she knows that's impossible right Now!)

***

Hang on, I have to quiet the "new puppy".

Vee insisted on sending over this stupid little robopet that she thinks will remind me of Melanie, our dog back on Old Earth. Nothing could possibly replace Melanie for me, just like nothing could possibly replace holding Vee in my arms.

Damn the woman!

What Vee doesn't understand (and what I guess women will never understand even though I spend a good deal of time explaining it to them), is that I have been programmed, trained, brainwashed, whatever you want to call it, to AVOID displaying any emotions I may feel.

This is my take on it.

A man does not go around sobbing, pissing and moaning about

his problems or things that have happened to him. He doesn't talk about how he feels, and he doesn't go around chattering about his sex life. A man doesn't need to.

You have your accomplishments inside your head. You walk around with them inside of you. It's what makes you stand tall against your enemies, against adversity. Hell, it's what gives you the juice just to stay alive in a killer competitive world.

Of Course, it's not all always about competition, sometimes it's a team effort. But if you're the boss, if you're the one at the top, people are going to be trying to shoot you out of the saddle. Unless you stand on your principles.

My point is, you don't let yourself show weakness and it bucks you up inside. It makes you feel as if you're not weak. It's like sympathetic magic, like working in a peak performance state; you have the energy to BE what you need to be.

I know it's been said (ad nauseum) that repressing emotions is bad for men and that we don't have a "full emotional life".

Well, for myself, and I can only speak for myself, I don't need this "full emotional life" that everyone thinks I'm devoid of.

I am happy. I am content. I am satisfied. If that is not what people are striving for, it should be.

On the few occasions when I've been extremely happy, I felt great and smiled. But I didn't run around screaming and yelling, or jumping up and down. I feel emotions inside.

And, I think I show them.

The misunderstanding is this: my mind is not occupied by emotional things; by relationships, by interactions with people, by what he said, what she said. I don't care. I really don't care.

Don't misinterpret me, I am NOT "hurting deep inside and hiding my emotions"; I'm simply unconcerned. Gossip, chit chat, jawing, and small talk, does not engage me. My mind is concerned solely with problems that are mechanical/electrical by nature, whose solutions are extremely practical.

Terrax is saddled with astronomical environmental and cultural problems, which I have volunteered to help solve. Now, it's my hell bound duty to help solve them, because billions of lives depend on the correct functioning of these decrepit, defunct, half dead life-support systems.

My MagicoMathematicians come up with computations and programming instructing the machinery on how to turn raw recovered materials into the few products that are necessary for survival.

The Big Four industries provide: electricity, clean water, purified air, sustenance, basic environmental shelter, enviro-protectants and simple entertainment. The WC administrates the Big Four, and the viewership provides endless entertainment in the forms of programming and participation on the Vidnet.

Vee's OTEC Techies keep the ocean energy generation systems on Terrax running properly, providing all the power supply for the life-support systems; these machines control the water and air purification systems.

The problem is how to repair failing machinery without metal. At present, the MOB controls almost all of the desperately needed supplies of raw materials recovered by their space cowboys, known as H.O.R.S.E.s.

Unbeknownst to Asna J., Vee and I have been privately funding expeditions to recover our own raw materials from the planet's interior.

We have gained access Planetside; both below the Warrens, with the help of the rebel group, the AGGIEs; and beneath the Pan Oceanic surface, our work being hidden from sight by the foreboding black oceans and the Northern OceanDome high above.

Our new source of raw materials is our deadly corporate secret, and I know that of all of the organizations, Asna's bloodthirsty crew would be the most willing to kill in order to obtain proof of it.

The Calava Corp (creators of the Prot-Shot) have their own rather disgusting, and deadly, I might add, company secrets regarding the ingredients of their raw product. Grandpa's Gang, as I call them, are the only producers of the sustenance necessary for survival, therefore their methods aren't questioned.

I don't pretend to be a "moral monitor", as the general viewer is so fond of calling itself, I leave that to the World Council, but one day those boys will be in deep shit if they are ever caught.

The MOB's Eedia MC (Electronic Entertainment Media Megalocorp) makes it patently obvious that it runs itself, having a monopoly on the one product that everyone here is addicted to.

On Terrax, without human interaction, this is a land of No's: No participatory sports, No social interaction, No real Food, and No physical sex!

That's why the MOB has the viewership by the balls, so to speak. What it boils down to is this: when all the basic necessities of life are provided; entertainment is the ultimate escape. The ultimate drug when there is absolutely nothing to do.

The MOB, the World Council and Grandpa's Gang were already here when all of us "new people" popped onto the scene.

In fact, to celebrate our difference from the people here, we dubbed ourselves Senselles, because we have a sensory memory cell, which once jacked-into the virtual reality net, can provide three dimensional programming for the viewership that blows 100% VR programming away.

Then, we realized the invaluable difference between us and them. With us, we actually experience our emotions, we feel our memories while remembering them. That's when we began to call them PHDs (Pseudo-Humanoid Drones.)

We cut a deal with the Powers That Be, that we would access the TDM regularly, to input our invaluable knowledge of another time, in exchange for a one third market share in the MOB's Eedia MC. Most of the Senselles went much further, promising VR programming and real time recording.

My MMs and I were the one demographic that decided to go renegade. Although we agreed to put a certain amount of our technical knowledge into the TDM, a wonderful and valuable experimental genealogical memory storage prototype developed by Grandpa's group, we did not agree to allow personal access to our memory cellular storyboards (known as memcell boards).

The TDM had been developed from a grand idea, a sweepingly compassionate concept. Grandpa, (that would be Laird's father), is very old school. He saw too many friends born to Terrax, in pain and suffering, saw what Radzeeze could do to a body. The general populace didn't really believe in mercy killing, because they thought that Euthanasia (the practice of taking someone else's life) was morally wrong.

So, Grandpa and his crew developed a machine that was to become a "savior to the sick". A sentient, conscious entity, one that would compassionately record the lives and experiences of those doomed to die and disappear from human memory.

The machine would allow you to submit to molecular disintegration and assimilation, technically called the Thermal Disimaging Machine, it was dubbed The Death Machine by the viewership.

Grandpa was understandably upset. He wanted people to realize that only by willingly submitting yourself to the machine, could it be activated. You had to be mentally, physically and spiritually in tune with the acceptance of your own demise, and be cognizant of what you were doing, for the machine to start.

In order that there be less public outcry, the first participants were terminally ill patients, who had volunteered to be disimaged and stored in the machine for their future generations to cherish. The Thermal Disimaging Machine would then painlessly imprint the sum of their lives; all of the complexity of their experiences, their memories, their thoughts and their knowledge into the machine, saving them for future generations.

These life experiences would be encoded and synthesized to combine the whole of human experience into the machine, along with every known fact from every form of media. All sorts of factual information, as well as imaginary worlds from motion pictures, televisions series, radio shows would be kept there.

In other words, the preservation of the full range of human experience, coupled with the full range of human knowledge and imagination.

I personally don't know why people think that it's worth preserving.

At any rate, the TDM had a two-fold purpose, obviously, it was intended to save the experience, the sensations, the minds and memories of all human beings. To record for future generations, so that all who came after would know what it was like to be human.

The TDM was created to save people from a terrible death, when they had a terminal illness no one could cure. This wasn't about the patient's families or some doctor deciding their time to die. The patient would voluntarily go into the machine, and if they weren't truly committed to the idea of dying; the machine would not disimage them. It would not activate. In this sense it was the world's first sentient, intuitive AI.

However, if they sincerely desire death, then they would be painlessly disimaged. It was as though they simply disappeared slowly, right in front of your face. It looked like they were being slowly pixilated, then dissipated into nothing.

Right away, people realized that a few of the participants didn't disimage (they didn't really want to die). The ones who were serious had taped goodbye messages to their families, and they did disimage.

Luckily for us, the practice caught on like mad, and quite a few people had been recorded into the machine when the Time Rips occurred.

By accessing that data timeline, Vee and I were able to figure out where some of the old machinery may have been buried in the earth's strata when we were transported through. It's hard to explain how, but in certain places under the earth's surface, physical objects of great value to us still remain buried by tons of soil, garbage, old buildings. The layers and layers of the detritus of human civilization.

We located some of the machines in the catacombs deep beneath the Melissae Hives, in the subterranean Warrens relegated as homes for the BayBays, the lowly worker bees of the planet.

Quite by accident, we discovered strata of mineral ore in the levels beneath the ocean floor. We were able to hide our work of delving into the oceans by the cover of the OceanDome, the enormous bubble that OTEC had used to encapsulate the power of the oceans. OTEC started rebuilding the OceanDome with living micro-organisms that were genetically engineered to create energy through a new biological process, called biovoltaics.

Before Vee and I were transported through the Time Rips, the MOB and the WC were afraid of the oceans, never venturing into the OceanDome. They didn't understand how ocean technology was actually providing a large part of the power for the planet.

Once we arrived, we started using the oceans wave power, as well as repairing the OTEC machinery (which takes advantage of the difference of the gradient in thermal temperatures to generate power and purify water for the entire Melissae Hive).

The last big player on Terrax is the Aerosuit Corporation, the makers of protective clothing for people, shielding us from harmful enviro-toxins. They have produced a spray-on and a paint-on Aerosuit of material that can protect a PHD's skin from the myriad of lethal chemicals that penetrate the Hives and the Warrens.

Upon arrival, Vee had refurbished their dilapidated Aerosuit Technologies Plant with her own personal funds, earning her a power position on their board of directors. She was the first of many Senselle replacements within their organization on the board.

Eventually, her top techie BayBays were running the place. (Smart woman.)

See, I did it again. Although I started by talking about emotions, I have to say that, as usual, my mind turned back to work. Between Vee and I, our job is to keep the machines humming that provide clean air, water and power for this dirtball we call home.

We directly provide basic necessities for the survival of everyone on the planet.

The MOB already had all the entertainment media sewn up

when we arrived, along with their aggressive launching of the HORSE space program. The HORSE program (Hydrogen Oxygen Recovery Specialists, Elite) is peopled by "space cowboys", whose special talent is "roping in" meteors for water reclamation.

They chase the bits and growlers, the bus sized chunks of ice at the tail-end of comets, the dangerously slivered debris of planetoids, all slicing through space at blinding speeds.

Some of the HORSEs even venture into using reverse space-time travel, traversing large patches of Pre-Rip Space. Bringing back raw materials like ores, even virgin metals, all desperately needed here to complete repairs on everything from space ships to the machinery for the air and water purification systems.

The Senselles.....well, they basically keep everyone vastly entertained in the NOW, the viewership able to anticipate a brighter future through escapism.

They are all my responsibility, like it or not. If only I knew how the MOB plans to influence the WC vote tonight. Then I'd know what my next move should be. I can't afford to take a chance on the outcome.

Guess it's time to put my dirty little cyberspies to work.

* * *

Frank/ie's Recuperation

I don't really remember exactly when I came back into my own consciousness, into an awareness of myself, or of my surroundings. The truth of the matter is, it wasn't that dramatic. Not like they show in the old reelies, when you simply open your eyes and see everything fully, and perfectly, like it was for the very first time.

It was more like drifting in and out of sleep, like fading from the place where you are awake, then slipping and sliding down into the dreamtime. You know, the dreamtime, that deep dream state where you feel nothing happening in the waking world, where your autonomic system is just rolling along on its own, your eyes moving with REM speed under your eyelids, your mind floating in the waters of your subconscious.

I felt I was slowly floating up and down, spasmodically able to feel sensations in only a few parts of my body, at odd times. At first I tried moving my legs, failing utterly. For awhile I just enjoyed the feeling of my head lolling gently from side to side. A little at a time, I was coming to the conclusion that I was in a semi-fluid, viscous material, not quite a solid.

I could sluggishly move through it, although it was very thick, it held me in place weightlessly, (this would allow my internal organs to regrow). It was neither hot nor cold – just a sort of warmish body temperature. It was soothing. It held me, comforted my battered body. And unbeknownst to me, it was revitalizing my organs.

I could flick open my eyes for a brief second, and was able to see only shapes and blobs. I began to be able to minutely move my appendages, which were gently floating around me, but the effort would exhaust me.

Then I'd slowly pass back down through the layers of consciousness, sinking back into the living jelly, returning to blessed sleep once again.

As the hours passed into cycles, my body and mind grew within "the mix". It nurtured my flesh, bones, sinews and muscles. I sensed more than knew that the mix contained an energy grid, infused with the DNA programming required to help my systems regenerate properly.

The program was to tell me who I was to become. But Auntie Vee's gen-ennges, in all their wisdom, could never have created what I became.

(What I became, was a Heeshe, an urban legend, a NowTime Myth. A freak.)

But when I first woke up, I didn't know anything about that; all I knew was that I was trying to recall what it felt like to float towards the surface of water, from underneath.

That was what it felt like to wake up into my own consciousness again. It's like what you feel after holding yourself underneath the surface of a pool, then let yourself float gently toward the surface.

The water parts around your face, and for a second your eyes can't focus on any one thing. It was that experience, of strong light shining directly into my eyes, that made me eager to sink back into the bio-sludge, and oblivion.

Oblivion was denied me.

Wad Medani and Auntie Vee had provided 'gramming from my (supposedly) favorite Senselle, Damaria. This 'gramme was designed to restart my artistic/creative/emotive module.

Her evening performance was up next on the hit parade.

* * *

the moment

there,
for a moment, just a moment
my heart softened. and I wanted
to reach out and cry with you, to hold you close again.
but the tear in your eye
was only a glimmer of light I saw there.
you changed the subject-
the moment passed.

By Kayla Raconteur

Damaria stood stock-still in the opening tribhangi (the thrice deflected posture), her limber body bent to precisely represent a DNA-helix.

This evening's exclusive performance of the sacred Odissi dance signified the willingness of the performer to allow the spirit of Ka to be infused into her.

The opening pose resembles a standing statue that will ever-so-slowly come to life.

Damaria demurely closed her eyes, but her lids weren't completely shut. Looking down at her own perfectly manicured, iridescent toenails, she saw her own reflection. She mentally admired her dainty ankles, strong calves and curvy legs. She was rounded and plump, rather than on the thin side. As it should be.

Continuing her critical self-inspection, she carefully considered her perfectly shaped arms, petite hands, bejeweled fingers.

She smiled to herself, outwardly not moving a muscle, as the music for the dance hadn't been cued.

Tonight was to be an advanced healing that Auntie Vee had specifically requested Damaria perform. Being a devotee of the Mahari tradition, her role as a devadasi required that she perform a number of healings every Wheel of the Year, so performing this dance would raise no suspicions.

Hers was a dance of exquisite mental and physical control, which she had painstakingly learned from her spiritual mentor and dance teacher, Mama Ish.

But, unlike the ancient craft of Olde Earth Odissi, hers wasn't the retelling of "The divine love tales of Radha and the cowherd God Krishna".

No, her dance told a story both simple and complex. It spoke of a girl from a poor suburban setting. Her father had taught her "take what people give you and consider them foolish for being generous". Her mother simply taught her to use her tits to get what she wanted.

Damaria had been taught that for a woman, "Sex is the only coin of the realm". And, as she had absorbed these lessons all too well, she was of poor moral character when she and Cami first met.

Cami was a wild child, willing to open her heart to Damaria, whom she immediately felt a deep kinship with. Cami treated her like a long-lost sister, and began to indoctrinate her into "The Ways of Perfect Love and Perfect Trust". She introduced Damaria to her spiritual circle, the Grove of the Amethyst Moon.

Cami bought Damaria new clothing, hoping it would help to get her a better job, and taught Damaria about budgeting and time management skills to help her increase her income.

Cami gave Damaria gifts beyond what anybody, including her own nuclear family, had ever given her. At least, that was what Damaria was putting out there. She constantly thanked Cami for her many kindnesses.

In return, because of the lessons her parents had ingrained into her, Damaria mistrusted Cami deeply. She was jealous of Cami's generosity and thought of Cami as weak, a sucker for being kind to her.

Oh, Damaria could play the game of perfect trust. She could smile and simper, and act as though she were humble, bowing her head, and batting her eyelashes at people.

And of course, she also pushed her tits at them. At least if they were male, that was her basic plan. If they were female, she would do all of the above, minus the flirtation with her tits.

Damaria had never learned to value strong women. During her rough and tumble childhood, Damaria's mother often abandoned her and her brother leaving them alone while she caroused with any hard-drinkin' man that happened by.

Eventually, Damaria learned that women are not important in and of themselves, but were only valued for the man that they could catch and use.

She also learned that you shouldn't ever trust women. They will hurt you and use you.

So, when Cami's unbelievable and unwarranted generosity was showered on her, Damaria did not know what to do with it. She basically grabbed up everything she could get with both hands, all the while plotting and planning how she would punish Cami for putting her in this situation.

Damaria couldn't forgive Cami for having this sway over her, because she knew she couldn't stop taking as long as Cami kept giving.

Cami loaned her the money for a car, when no one else would. She gave her a bed to sleep on, furniture for her apartment; even fronted her living expenses. She treated Damaria like the little sister that she never had.

All this flashed through Damaria's mind as the music was

about to strike the first note. The music would signal her body to release her frozen pose, one muscle at a time, slowly beginning the motions of the Dance of the Rotiarte, the dance the Odissi Grande Dames had named for her.

This was Damaria's favorite dance (quite an inside joke for her). She was perfecting it, secretly living for the day when she would triumphantly dance it over Cami's dead and wasted body.

But, until that moment, no one must know of her intent.

HAH! Cami would never see the betrayal coming!

Suddenly, all thought disappeared from her head as she heard her musical cue, and the madal player strummed softly. As the tabla took up the beat, the bansuri flutist waited for his cue.

The performance had begun.

She began to dance, completely unaware of her surroundings.

Beginning with the Nritta, she spun in complex, ornamental patterns, using her body movements to place the story in space and time.

Breathlessly ecstatic, she easily moved into her Abhinaya, the ritualized mudras (hand positions) and facial expressions used to describe the emotions of the play participants.

Shifting her torso from side to side, she hypnotized Laird and Randi, reciting two ashtapadis (entrapment poems consisting of eight couplets).

By the end of the dance, she would able to instill into him the Command That Could Not Be Denied, just as Auntie Vee had instructed her to do.

All they needed to hear to execute the command, was the recitation of "The Emphatic Invocation of Jayadeva", which only Auntie Vee and Damaria knew in full.

Damaria's Dance ended.

The world remembered to draw breath once again.

Things had changed, and maybe not for the better.

* * *

Freak of Nature

I can't blame anyone for my own personal hell,
I'm holding my own feet to the fire,
I'm steering myself off-course,
I'm misguiding my own star.
And I can't stop myself,
cuz it's gone on far too long.
And I can't stop myself
cuz it's gone on far too long.

-Song By Kayla Raconteur

Barrage Rhyzin, an up-and-comer in the MOB, although determined to avoid the Sextel Channel 'gramming, had been inadvertently jacked-in to Damaria's Dance (from her point of view!)

He witnessed the beautiful, lush Senselle Randi, cavort and play with her meshling, Laird. Without knowing how it happened, Barrage realized he was lusting mightily for the strumpet. He was angry that this ultimate invasion of his privacy had occurred. He'd never had anyone manually override his viewing preferences before and he didn't know what to make of it.

Just as he was about to place a call to his "boss", he was peeped at the highest security level available, and was instructed to Withhold Judgment until a meeting between them could be arranged.

Wait a nano! Asna Janghaghir had never before met with Rhyzin in the flesh.

Now completely confused, Barrage coolly ripped out his jack-in and remained UTL for nearly a half hour real time.

He knew that would upset the bastards. But he had made a command decision to let Asna Janghaghir have a taste of his own medicine.

Let him know what it feels like to be jacked around.

* * *

Frankie is Dead, Long Live "Frank"???

I am reborn, with only a few minor changes.

Now, I'm equipped as "Frank"!

Having taken one-quarter Turn of the Wheel to become fully restored in Auntie Vee's Uberlabs, and having been subjected to every scrutiny and indignity a PHD can bear, I have been officially "released" to "go on with my life".

Thank you so much.

"My life" is an OUT and OUT joke. The filths.

I know, I know, they have their secret reasons, their grand experiment. Yeah, they've tried to tell me all about it. I don't care.

What it boils down to now, is this. The only thing I want to do with "my life" at this point, is to kill "The Milliner". But, how? I'm no one, not so much as a blip on the net.

I have no past, no family and no chedits to my name. I know I can't possibly locate The Milliner alone, not with 27 billion people housed above and below the surface of Terrax.

At least, not without the help of the MOB (the Media Overlord Barons), because they control all the information on the planet.

The good old MOB. Think of them as descendents of Ted Turner, the great media mogul from the 21st century who gobbled up a lion's share of the major media companies and built them into mega corporations. Anyways, that's all ancient history.

Now, their main man is the MOBster who calls himself Asna Janghaghir, (which means "The Great Fear"). He's the brainiac who patented most all the World Council's standard issue products, like: instantly ingested foods (protshots), spray-on clothes (Aerosuits), mobile-modular home units (cubbies) and electronic entertainment media (Eedia). In other words, he controls a monopoly on every service or product required for survival on Terrax!

His factories were the first to pioneer the proteo-injection system (dubbed prot-shots); and we all know that "every hot-bot needs their prot-shot... daily".

This was how the MOB became the foremost power in the known universes, the World Council notwithstanding. And this was why, when the Senselles showed up and broke that monopoly, the MOB instantly became their bitter enemies.

I figure that, somehow, I have to get a job with the MOB, because they catalogue (jack directly into) every living entity on the planet. They will know where The Milliner is, because every lousy PHD drawing down their share of hydrogen has to be registered with the MOB in order to earn chedits.

As far as I know, no one can escape detection from the MOB's grid.

I realize the irony of having to beg a job off the people who probably arranged to have my family killed in the first place, but there's nothing for it. They are the only ones with the power, the tools to find him, so I need an entry level position, like a.....a... Media Information Specialist.

Yeah, an MIS! That's it, I'll get a small time gig, and quickly elevate through the levels. I'll start as an entry level security specialist, then go on to the assignment of MIS, where I can tap into the information I need to find him.

I have to talk to Grandpa about getting me the job. I know he doesn't condone how I've been salvaged and then remade, just to test a new operational theory.

I mean, he knows all about Wad Medani's involvement, what with his sodding MagicoMathematicians running around reconstructing me from a valise of human DNA for his own selfish purposes.

Gramps knew it was wrong, but there's nothing he could've done about it. At any rate, I'm here now, and I should have a say in my own life.

(Yeah, it's true they re-grew me in a proteo-solution injected with my original DNA that they recovered from a dark alley in Suiterville. Not the chronic start I'd have wished for, but it will have to do.)

In regrowing me from the original implant, the MMs discovered they could disimage a person's body (take it apart at the molecular level, basically dismantle it while retaining all the unique mental pattern and information). They accomplished this by putting my deteriorating DNA into the TDM (Thermal Disimaging Machine), and bringing it back from the brink of death with the Bradford Reverse Osmosis Treatment.

Get it...they BROT me back from the dead. Eeeeeeeeeeeeewwwwwww.

They did use me in a way, but I was dead meat when Cami "found" me anyway, so I'm not sure who owes whom a favor at this point.

Dayyyyuuum.

I hate all this philosophical bullshit. All I know is that the only time I want to be alive is when I begin to think of how I'll kill The Milliner. To that end, I need to start my mission with the help of my only ally, Grandpa, Wad Medani's original mentor.

At first, Gramps was less than receptive to the idea. Then he slowly began to let himself be convinced to go along with my plan. I should have known that in itself was enough to make it suspect, but I was so eager to go and kill The Milliner that I was reckless, and didn't think it through.

* * *

GRANDPA

As for me, I knew what Wad Medani and Auntie Vee were planning to attempt with this hard won knowledge, but I still don't quite understand why they're doing it. The outcome they desire is uncertain at best, and could be even more disastrous if it succeeds than if it fails.

I find it distasteful that the MMs found this person in a bag, took those remains and recreated a new person from leftovers. Even if it would help to know how to bring a person back from the dead, I don't see the point.

I believe, each being has an amount of life bequeathed them when they are born, call it destiny or what have you, and they cannot cheat that clock.

But Auntie Vee and Wad Medani are convinced that if they can disimage and then reimage themselves here in this world and in this time, that they can be in each other's physical presence once again.

Grandpa sighed, then took a deep slow breath, hoping the extra air to his blood supply would invigorate his brain. He strained his old worn-out memory, trying to remember the feeling of loving someone that much, so much that you would risk death to hold them one last time.

He utterly failed. He truly had never loved anyone that much and did not even remotely believe that it was possible. But this is what they claimed, so he had to believe it was true, if only for them.

And now, after all that drama and risk and high science and high handedness, this is what the new being they've risked so much to create wants to do with its new life. All that planning and effort, just to be thrown away on this useless vendetta.

As I stare right into Frank's eyes, I try to fathom his being, his soul.

I am trying to sense if this creature has a soul.

Is it hubris to believe mere PHDs can infuse matter with spirit through mechanical means?

Again, I bring my mind back to the present, trying to understand Frank's need for revenge. It seems Frank is focused on that one thing, the need to kill The Milliner. To exact revenge for destroying his life and all the good that was in it, for having changed Frank/ie's life irrevocably, and certainly not for the better.

Now Frank/ie had no one, no single person to care for him, only a bunch of mad scientists (who viewed him as no more than a lab rat), and an unsavory collection of power players who were each interested in him for their own shady reasons. From the biggies (Auntie Vee, Wad Medani and Cami), to the techies, and possibly even the MOB, everyone wants a piece of Frank.

Poor bastard.

Grandpa felt a guilty twinge or two of his own, for he had helped them with the reconstruction of the blasted BROT machine.

He had reconfigured the TDM in order to facilitate the genotypic molecular-structure regrowth cycle. He didn't trust Wad's MMs to take apart his beloved machine without destroying it, so he assisted with the reconstruction.

Now Frankie stood before him, ready to throw his new life away. Gramps felt an almost proprietary right over her/him (damn, this is confusing). Like a futuristic Dr. Frankenstein, who felt he owed his monster some sort of protection once he had given it life. He deeply regretted that the creation he'd helped to make would be so lightly wasted.

Dammit, Frank is alive, an independent living being, with his own will and desires. And there is nothing I, Wad Medani, Auntie Vee or any of them can do about it.

As I look into Frank's eyes once again, I realize one more thing. I owe Frank something, too, and rather than let him do this thing on his own, I will help him. To a point.

I began to ponder my varied and sundry connections; and suddenly begin to think ahead of the game. I decide to go ahead and help Frank get a top level MIS position with the MOB, so that he can start tracking The Milliner as soon as possible.

I think that this way, I can keep an eye on Frank and see that he stays out of trouble. And hopefully, before he makes his move to kill The Milliner, I can talk sense into him, and stop him from doing it.

I'm not sure if it's the right thing to do, to get involved in this mess. But I'm so damned frustrated with all these people right now, I just want to absent myself from all of them; Wad, Auntie Vee, Cami, Frank/ie – the whole lot.

I have my own problems going on with the World Council and my son, Laird, so I don't want to spend too much more time on this madness.

Let it Be.

* * *

Frank's first assignment with the MOB, as an entry level MIS, was to track energy pattern anomalies within the Vidnet system, charting any unusual occurrences and data bleeps. This was how I was able to start determining the criminal element within the viewership, bringing me one step closer to finding The Milliner.

In the meantime, it's time for me to jack-in to my favorite two rapscallions, Cami and Cherry, as they continue their latest adventure, entitled "Mardi Gras Done the Hard Way, Part 2".

* * *

An Epitaph of Idiots

The next day Bruce and Don took us safely over the border into Arizona, gave us a couple of bucks and dropped us off at a McDonald's.

Cherry and I cleaned up the best we could, which was limited to washing our faces and finger-combing the hair out of our eyes.

With at least another two days to hitchhike we had to look as "harmless", "innocent" and "non-disgusting" as possible, in order to make it out of the state in time.

We grabbed some breakfast and headed on down the road.

I regretted not grabbing a toothbrush before leaving home. Instead I bummed a ciggie off my buddy Cherry. I had my own smokes, I just bummed from her cuz it drove her crazy, and, cuz' I got away with it.

I held the map up in the air, trying to get the folds out of it. Squinting my unshaded eyes against the early morning light, I looked speculatively down the street.

"Are we set?", she asked, itching to be off, to be anywhere but this no-place Podunk little desert town.

She's always cranky in the mornings, and we'd only been able to get one cup of coffee each. Funds were running dangerously low.

Shit!

This run-away stuff was trickier than I thought. Between trying to keep us fed, in smokes, not frozen, and not raped, it was a constant 100%-wide-awake effort.

The up-side to all this is; I'm free, I'm making all my own decisions, paying the price when I'm wrong, and glorifying in surviving when I'm right.

And, Hey! No one had slugged me since 4pm Thursday afternoon- life was good!

We heard a car goin by, stereo playing our road song, "Truckin', like the Doo-Dah Mannnnn", and we both broke out in a huge identical grin.

We thumbed a ride to the nearest freeway on-ramp, trying to be extra careful because the APB my dad put out on us was probably in effect from California to the border of Canada and east as far as the Texas border.

(I knew this from previous experience as a run-away. Unfortunately, my uncle works as the warden of the state penitentiary in Arizona, the real badge- carrying, baton-swingin', gun-totin' thing).

My male relatives from Arizona detested the fact that I used "big words"; that I lacked the proper respect for rules; that I didn't know a woman's place, blah, blah, blah.

So, they are always sure to be instrumental in helping to drag me back to parental authority, and punishment.

For now, they are content with what my parents dish out, as long as I'm a minor, and out of their reach.

Once I'm eighteen, they have more diabolical plans in mind, as they take no uncertain pleasure in reminding me.

One cousin in particular, L.G., loved to loom over me threateningly, and tell me "Someday, little girl, I'm gonna put your ass in jail!"

I would ritually respond by intoning, "You're gonna have to catch me first!"

As for the present, thus far we've evaded the state and local police in two states, and I'm feeling nigh invincible.

Unfortunately, right now we're beginning to have the inverse problem of last night. Instead of being freezing, it's getting hot.

Hotter by the minute, it seems.

Standing at the abandoned on-ramp of a forgotten highway in a small town in the middle of the desert, we belatedly realize we've made yet another bad judgment call.

The only shade provided in this hell hole, is a sign on the highway, which reads: NEXT TOWN >>>>>>>>>> 90 miles!

YIPES!!!! We hustled on over to the back side of the sign, which was the shady part.

We figured it would provide shade for at least for another hour.

Actually it appeared to be shrinking by the minute, when suddenly we noticed....... the damn names.

All these names, etched into the back of the miserable sign, written with pencil, pen, scratched in with coins and rocks.

An epitaph to idiots, it is a long list of losers who have been stuck here before us, engraved on this monument to egregious errors.

Sigh.

Cherry can cuss, when riled. She fired off a stream of invective and gave me a hard look.

Shit! Shit! Shit!

THE PERFECT GENTLEMAN

Taking another look around this vast stretch of sand, cactus, lizards and more sand, the view went on for miles in every direction.

I noted, no skeletons around the immediate area, and told Cherry it proved that at least someone had gotten away from here.

She replied, "that only proves their remains have been picked clean, etc., etc."

All in all, I was very glad I didn't have a watch.

I don't know what would have made me crazier, to know we had been out there in the blazing heat for hours, or to know it had only been a few minutes that seemed like hours.

I was beginning to miss last night's bracing temperatures, when a wonderful man in an air conditioned car pulled over, and waved us in.

We were so desperate to get outta there, we'd have gratefully gone with Jack-the-Ripper; and piled into the front seat of his car.

The backseat had a bar across the top that had suits hanging from it. He was packed to be on the road for awhile, from the looks of it.

I directed one of the air conditioning vents right at my face and sat luxuriating in the feeling. Cherry followed suit.

Greg introduced himself, and told us he was on his way to a town a couple hours south of LV, where we were heading. He offered to take us the rest of the way. Turns out he was an alarm clock salesman, on his way to a seminar of some sort.

We tried our bullshit story out on him, the version we would tell the pigs if we got caught.

He looked at Cherry's dirty, messed-up hair and clothes, and ran an eye down my equally disheveled appearance.

"Right" he said shortly, and with a dubious snort, left it at that.

It was obvious we were lying, and that he knew it. He probably had figured out the truth by then, that we were two under-aged runaways, but he was a gentleman about it. First class.

In fact, that evening, as we pulled into the town, he offered to buy us dinner at a fast food joint, his treat. We went to an all-you-can-eat buffet and chowed down. I hadn't felt full in days and stuffed myself, including dessert. Cherry ate like a wolf too.

Over sundaes piled with whipped cream and nuts, he casually asked if we had a place to stay the night. Right away I admitted we didn't.

Greg asked us if we'd like to crash on the floor of his hotel room, and promised us he wouldn't try anything.

Call me crazy, but I'd already kinda vibed him out, and felt he was telling us the truth. I believed him, and gave Cherry a long serious look.

She might not say much, but Cherry was psychic, in her way,

and she nodded a brief agreement.

As always the front man, I agreed we'd love to take a shower and get a decent night's rest.

We drove back to the hotel. The huge king size bed in his room was big enough for the three of us! We took turns in the shower, scraping off the scum of the road. Then while Greg took his turn in the shower, we made a game plan to get to LV by tomorrow afternoon.

After showers and what was left of the late, late night movie, we slept like the dead. True to his word, Greg never tried to fondle us or come on to us in any way.

He still shines in my mind as the ultimate example of how good people can be, without wanting anything in return. He was a life-saver and the perfect gentleman.

We parted ways Saturday morning, and even Cherry was a bit misty-eyed as he dropped us off at the off ramp.

Looking at each other, we couldn't believe we made it this far! We were in New Mexico now, heading north to our long sought destination.

The rest of the ride would promise to be much rougher.

* * *

2

Who Rules the M.O.B.?

A Pale Shade of Red, circa O.E. 5020

once i was hidden away,
sneaking through brown hills in summer,
gliding from behind a rock's shadow,
to stand on parched grass,
snapping and cracking beneath my paws.

my eyes gaze upon a vast expanse of gray-white,
long as the day, it stretches on
from the beginning of light
to the end of the world in the other direction.

i see it
and
i fear it.

- BY THE BARD, PHOENIX GAET

* * *

Who Rules The MOB?

Asna Janghaghir, whose very name means "The Great Fear", sat comfortably ensconced in his thickly padded, emerald green egg-throne. He felt imperturbably serene, surrounded by luxury.

He was observing the antics of his newest toy, a gene-modified clone of Seleste's favorite feline, Isabel, a multi-colored Calico cat.

Seleste had once told Asna that the name Isabel (whose origin hails from an ancient Gallic people) refers to the mottled coloring of the Calico. Asna satisfied all his petty desires, often gifting himself many such expensive items.

He loved the tactile sensation of the physical objects he owned, each of them rare and costly. While stroking the soft, downy fur of his GenMo Cat, or feeling its sandpaper rough tongue licking his hand, he took the deepest pleasure in his "sense of touch".

He even had a perversion for "eating" real food. He often commanded his underground connections to bring him samples of The Cook Nynah's food. No matter how sick it made him, he endured the pain of ingesting solid food because he loved the pleasure of tasting it.

You see, Asna was a cybernetic creature, one who had evolved through countless annuations of genetic engineering surgery.

Over time, the gen-ennges had toyed with the idea of creating a new breed of PHD. One who was not fully a Senselle, yet who could be bred to feel a small range of emotions.

Of course, whether physical feeling (sensation) constitutes actual emotion, is still a raging scientific debate, one which Asna's own personal research will greatly influence.

But Asna cared not for the pinheads and their assessments of what emotions were. He knew what emotions were, because He *felt* them.

Asna could (of course) see and hear, but then all PHDs could do that much.

What differentiated Him from them was that Asna could also smell and taste. And, His sense of touch was heightened to a sensual data input level, unlike the groveling masses.

True, His sense of touch wasn't active all the time. Like a fond memory or the recollection of a dream, it would come and go in gossamer bits and pieces.

Unfortunately for his enemies, Asna was kept in a state of deep, perpetual self-awareness, due to this onslaught of sensual data input. He caught nuances in their conversation, tone of voice and body language that came his way, and savored them like a sudden breeze on a hot afternoon or the touch of a lover.

The wealth of information these sensations brought to him was indispensable to his ability to second guess his opponents.

This morning-cycle, Asna was not fully engaged in his toys, so much as he was relishing the wonderful opportunity that had fallen into his rather ample lap.

* * *

Today is the day-cycle I have waited for, for an age and an age. I have waited ever since that day long ago, when Auntie Vee and Wad Medani burst onto the scene and stole my thunder. They undermined my monopoly, my wealth and my death-grip stranglehold on Terraxian resources.

That day they created their own nemesis.

While attempting to forge this "unbreakable alliance" between their Techies, the BayBays, the Senselles, and Wad's insane MMs; to vie against me (The MOB) and the existing politico-socio-structure (The World's Council and Grandpa's Cronies), they've sadly underestimated us.

I knew their little alliance could not be broken from the outside, as with a siege. I have pondered the dilemma for gigannuations, and resolved that the only way to break them is to divert the Senselles away from their own internal 'gramming.

But how?

I have thought long and hard on it. I've been waiting for just the right opportunity. I finally understand that only betrayal from the inside can truly destroy them.

Just as "bad brandy" (an insignificant amount of bad wine that's mixed in with a huge amount of good wine) can taint the whole batch; one traitorous Senselle will bring them all down.

It's quite delicious that I found that miniscule bit of bad blood within the Senselles – someone who has the perfect genetic moldings to be a traitor to her own kind, and to be able to hide it flawlessly.

As a double agent, I know that she'll be invaluable to me. Why do I not name her? Even now, in my own private thoughts and subconscious mind, I will not reveal her name.

I am well aware there may be a mole in my organization, whether serendipitous or planted, I sense a fatal weakness here, as there are in all genetic organisms.

My organization was created to exist as a sentient genetic structure – composed of a strong double helix.

Barrage Rhyzin does not know it, but he is my opposing strand, the intertwined and infallible back bone of The MOB. We twirl round and round each other in a syncopated dance, at perfect odds and yet in perfect balance with each other. Opposite, opposing, contrasting and clashing, yet bound together, eternally spiraling outward. We drive the Wheel of the Year forward.

He is genetically modified to be who he is, yet there is also a generous handful of chaos factored into him, some random unknowns.

Barrage also happens to be my beloved nestling. I created him, with the help of Wad's "Procreatrix", the code name assigned to the seven top gen-ennges who developed his genetic modifications. They, as well as many other players in this saga, will remain unnamed; however, I do know who was responsible for introducing the "surprises" into my son's DNA mix.

He will be repaid in kind, in time. But, I wander. I digress.

Ahhhh.... the exquisite nature of the Senselle betrayal to come is that it will come from deep within their organization. Tis' that old misplaced trust that's a bad move, strategically. They didn't take the time to properly research this particular Sister (as they like to refer to themselves).

The Senselles, who have always been meticulous in screening their applicants, let this one in on the most preposterous sob story you can ever imagine.

But, then again they are emotional creatures, drenched in their own imagined feelings. They entirely miss the fact that they themselves are projecting this supposed emotion. When in fact, the "emotions" we ascribe to others originates from within ourselves, then we compound our error by mistakenly thinking that other people are returning those emotions.

At any rate, I now have several informant-links into their precious Sisterhood.

The other prong of my attack on the Senselles is to infiltrate the AGGIEs. I have a “plant” in their organization looking after another angle, a certain Pavarti, the supposed head of the Above Grounders.

It came to my attention that Pavarti plans to interface a particularly nasty UVP with the Vidnet viewership. I am familiar with this particular strain of chaos, and although I won’t go so far as to say I employ “the monster“, I know that Pavarti plans to unleash it on the unthinking populace.

Coming from the MOB’s Labs, I know the creature is a fine specimen of UVP (Ultra-Violent Programming), someone the viewership will not forget in many annuations.

They will call him “The Milliner”. He has an incredible potential RQ and an embarrassingly large Contemplated Value for wreaking his unique brand of chaos. He’s an evil side-show attraction that I can fully utilize.

Later.

I’ve always thought that a taste of UVP a day is necessary to soothe the wild beast in all of us. Especially when a ‘gram appears be improvised, without the victim knowing they’re being taped, which is a true rarity.

If you ask me, the AGGIEs will not be able to control The Milliner, once he is activated. But the AGGIEs do not take advice from me. It is absolutely necessary that Pavarti believes he is acting out his own free will.

Unbeknownst to him, my dear Pavarti has been laced with all sorts of egoistic tendencies and addictive vices. We made sure that sex, in particular, was a weakness in him. We knew when and where to catch him in a vulnerable state, and had him hard-jacked directly into a private session with......my little informant, Benedictine.

As a Soyal gift from the Senselles to the WC, Benedictine, a Temple of Isis priestess, is being sent to dance for the World’s Council.

Pavarti, imagining himself a part of the power struggle here on Terrax, habitually tunes in to The Political Channel. I made sure that Pavarti had "secret access" to several behind-closed-door meetings, which were being recorded for his benefit alone. I fed him the information I wanted him to know.

It was almost too easy.

Pavarti, falsely filled with his own self-importance, is now hooked on the comcasts. I know I can jack him in any direction I want, and I want to let him dangle in the wind, panting like a dog after my Benedictine. He wants another taste of that performance. A much larger taste.

He should take care with what he wishes.

Later this eventide the World's Council plan to take Laird to task, because lately there's been something off with the balance in his decision making. There've even been info-reports leaking out that his relationship with his meshling is monogamous (which is strictly forbidden by World's Council statutes).

The WC came to a General Decision that Laird and Randi, his meshling, must participate in a Temple healing. The session will be facilitated by an authorized Senselle Temple Priestess, and will be recorded by the TDM (proof to the WC that Laird is back on track.)

Pavarti, having tapped into this information, decided that he would jack-in to the public portion of the sexual healing. Knowing that he has this particular weakness, my heart flutters for a moment.

What if the little perv gets hooked on the wrong Senselle?

Ah, the unforeseen details are always what catch you up!

In studying his profile, I find the odds are that Pavarti will desire them all, both the Senselle women, and Laird. Ultimately, I believe he'll be too intimidated by the stronger woman, Randi, and will leave her to Laird.

My sweet liar, Benedictine, will appeal to his sense of insecurity, his own inner darkness. I will see to it that she personally gets to meet him at the next "clandestine" AGGIE meeting.

They will talk together apathetically about rebellion, will

complain that "the WC sucks air" and that "the MOB is ripping off the viewership". Then they'll congratulate themselves for being better than the average Jack, and like most rabble rousers, once again ignore the fact that the most important step to change is action.

Pavarti, hoping to impress B., will offer her a duel mod jack-in, sharing one of Cami's adventures together (subconsciously sending B. into a jealous rage). This will only serve to fuel B's anger against her Senselle sisters.

Pavarti will discuss the triad of power, and eventually, as is his wandering vocal manner, he will turn to the inner workings of the Senselles.

Benedictine, knowing her knowledge superior to his, will share just enough information to hook him, sending him into a swoon. The ultimate game of desire and betrayal will begin.

I have found a way to turn their private plans of freedom and double dealing against them. Whether he knows it yet or not, Pavarti is only a little tool of mine – a toy that I use to amuse myself with – something to create a stir on the Vidnet. A mere distraction.

The viewership finds rebellion endlessly fascinating. So I'm simply providing fodder for their greedy, simple minds. Little do they know how we plan their every move, and then supply the very need we create!

The AGGIE's power isn't real; they don't have a chance in a gigallion of becoming an independently funded organization on their own. Their entire premise is ridiculous; one cannot live on the OUTside without protective measures. The environmental toxins are enough to defeat mind, flesh, and machine.

Only by the Grace of the Melissae do we exist. The Hive itself is made of GMO biophotocells that the Melissae engineered in their labs. It's composed of an organic material that reproduces and maintains itself. The organic material collects variant star matter and translates it into the energy we need to power the machines that purify our air and water supplies.

Most of what the Big Four Industries can produce, our MOB

factories can also produce. I will prove to the viewership that they don't need the Senselles or the power hungry MMs anymore.

In my hand, I hold the end of one single thread that I can tug on.

I have time on my side. With help, I will eventually unravel their entire alliance.

Being patient, I placidly await the day when I will see their smug, complacent sense of safety being brought down around their knees.

Once again, I will dominate the leadership of the Hives and the Warrens, all will submit to my every whim.

It's what I live for. And, since I have reprogrammed my own DNA, it's the only thing that I will ever want to do, the one goal I will work tirelessly to accomplish.

That's my basic strategy. My Benedictine will infect not only the Senselles, but my little AGGIE leader. At that point in time, I know he will do anything to have her.

I can arrange to have her submit herself to him completely and then withdraw. Ahhhh....denial and appeasement, the worst and the best pleasure game of all.

Then, having infected Benedictine with DNA that will turn her into an addict for Pavarti, I will have them both, unwittingly, under my control. The three of us will methodically destroy the high and mighty Wad Medani and the noxiously overconfident Auntie Vee.

Speak of the devil, and up he pops! I have Wad Medani on an incoming peep. What could be abrewin' now?

Whatever it is, it must wait.

I never miss an e-pisode of Cookin' Wit Nynah! To jack with Wad, I'll peep him back later.

My favorite gramme is about to start.

* * *

Cookin Wit Nynah!

Nynah the Cook was a comfortable looking, 65 year old

woman (Olde Earth years). She was 5'5", plump, as opposed to fat, and weighing in at about 150 pounds, she was a solid, healthy middle-aged matron. With olive-toned skin and freckles, she has muscular arms, because she's always lifting 10 pound bags of flour, and hefting around pots filled with tamales and frijoles.

An excellent cook, she can make all sorts of food; from simple American fare like steak, potatoes, and vegetables; to inventive tossed salads, then top it all off with an exquisite dessert. She improvises meals of flavorful ethnic food, cooking Mexican food with zing, Italian food with flair and can easily whip up a mean chop suey!

Her cozy kitchen has the heavenly smell of freshly baked breads and cookies wafting through, as well as her luscious fruit pies and sweet pastries.

Just as Asna tuned in, he saw The Cook bustling around her kitchen making a light breakfast. It was relaxing for Asna to watch her; to completely immerse himself in the show, seeing what she saw, smelling the aromas and anticipating the flavors, leaving his own troubled world behind.

Nynah was whistling some unknown tune, she felt good this morning. Up at her usual time, around 5:30 am, she wrapped an old robe around her ample middle and puttered off to put on a pot of coffee before she did anything else. She doesn't like her coffee strong, so she uses the little measuring cup that comes right in the coffee can and scoops up a couple of heaping capfuls for every 10 cups of water. She knows it will be perfectly tasty and wonderful.

Asna knows The Cook prefers regular French roast, but will occasionally sprinkle in a dollop of special coffee grounds, maybe some chocolate almond, or her personal favorite, mint chocolate. She loves the dark chocolate too. Sometimes she sprinkles in a dash of cinnamon over the coffee grounds before they begin to drip through, to cut the bitterness.

While the coffee was brewing, which took about 10 minutes; she took out a frying pan and sliced off a bit of real butter (she liked to use stick butter, because it's easier to accurately measure).

Putting a dab in the bottom of the pan, she set the burner on a low flame, making sure that the butter was completely melted. Scrambling in a couple of eggs, she sprinkled in salt, pepper, and crushed garlic, finishing with a healthy dash of Tabasco sauce.

She liked to scramble eggs dry, so she would stand there daydreaming as she slowly stirred them. She popped a couple of pieces of wheat toast into the toaster, and got out the tomato jam she'd just made a couple of days ago.

Scooting the eggs to one side of the pan, she took out two strips of bacon and started cooking them next to the eggs.

Carefully choosing a breakfast plate with the delicate blue flowers on it that matched her blue speckled coffee pot, she poured herself a cup of strong black coffee.

Retrieving her medium light toast out of the toaster, she adorned her plate of eggs and bacon with toast.

She sat down to her simple meal, with her homemade tomato jam on toast. Heavenly.

As she slowly consumed her meal, Asna "virtually shared" the meal with her, savoring the salty flavor of bacon and eggs. The Cook had put just a tiny bit of shredded cheese on top of them, making them so delicious. Taaaaas-teee.

Nynah was remembering breakfasts like this spent as a child in her mom's kitchen. It was a fond memory, and she recollected it in the privacy of her inner mind. She held the memory for as long as possible, then her show was over.

* * *

Feeling somewhat refreshed (and even a bit guilty) for having taken a break, Asna turned his mind back to his sixteen incoming calls, his hundred and forty four bit Vidscreen inputs and his recommunication with the MOB, which had been temporarily disconnected while he was jacked into The Cook's breakfast.

He peeped Wad, with a slight sense of anticipation, mixed with wariness. What in the world could the great Wad Medani want from him?

* * *

Soyal...The Mid-Wheel of the Year Celebration

This is Wad Medani signing off from the OTEC Channel, where we are "Fueling Your Future."

Wad Medani took a deep breath.

Today I feel my years.

After checking in with Vee this morning, I found that all is well. Apparently Cami's retrieval mission was successful; they now have "the body" onsite at Vee's private labs. I needn't spare the energy worrying about what's going on in the Melissae Hive, Vee's got the project covered.

Knowing that the best personnel is available, from my top MMs, to Vee's OTEC Panatechies, and that we've some of the sharpest hardware people around working on the project. I know I can leave the Big Experiment in their capable hands.

I have another mission to accomplish.

You know, I'm a physical guy; it's part of who I am. I'm of Viking ancestry; my people were hardy and strong. My dad worked 16 hour days until he dropped into his grave at age 90. That was Olde Earth years, when there wasn't nanotechnology and cyber-med to regenerate organs failing with old age. You had to play the hand you were dealt.

Yeah, the old man – he did it the hard way. I hadn't thought about him in eons, but today, he came to mind because of the dirty, physically grinding work ahead.

Unfortunately, I couldn't let Vee know that I was going into danger, not this time. She just doesn't understand why I have to take these chances. Particularly when she and I are so close to a personal breakthrough.

As it stands now, we can only enjoy each other's company via some form of virtual reality holographs or holograms (which seems like a cheap bastardization of the real thing.) It's one thing to be inside someone's mind; it's another thing to be in their physical presence.

I used to love it when she would laugh and come running up to give me a big hug. I felt like I had the whole world in my arms. You know it's strange for me to be so... romantic....emotional...

on such an important morning. I've just been thinking about her a lot lately.

Damn this project we've got going. Will we really be "us" after we are regenerated? Will we have the same memories and feelings for each other? Will it even work? Can we survive the procedures?

What will happen when we're gone? There's so much unanswered, with only the word of da Silva that it will succeed.

But I can't afford to let any of that distract me right now.

Today, I have a particularly sticky situation to take care of. In the Southern OTEC OceanDome, we have several ROVs operating in that mess that we call "the ocean".

Technically it's not an "ocean" at all. It's more like – like a huge battery. A living fuel cell that generates and stores the energy we need to keep the planet's computers and machinery alive.

The other half of the hybrid energy generation system lies above the surface of the planet, known as the Hives; which are composed of organic, living polycrystalline matter that generates electrical energy.

We keep this place going with our nanotech energy intensifiers; keep the air purified and the OTEC Underwater Tidal Fence Division running. Everything has to be maintained and repaired from Wheel of the Year to Wheel of the Year.

Everything usually runs like clockwork. Until yesterday, when my techies in the Southern Hemisphere Dome were power lifting a ROV out of the water for routine servicing. Suddenly, their power indicators started going off the chain.

Even though the diagnostics tests they ran didn't find anything wrong, the ROV was transmitting a weird energy signature code, similar to phantom energy (P-energy).

The ROV operations supervisor on duty, Arpad Reitan, immediately peeped me, because without a protocol for Planetside P-energy containment, he didn't know what the hell to do.

Vehicles from deep space are currently the only craft known to suffer exposure from P-energy, and this little ROV had never been out of the ocean's waters, much less in deep space!

After careful consideration, I decided that I had to call in Asna (I will call him by no other name and acknowledge no formal title).

I asked him to send his Eedia cyborgs to the S.H. OceanDome to assist in decoding the ROV's energy signature in our presence, so there would be no shenanigans.

Although his Eedia cyborgs have personal disteldt stealth shields around them to block us from their incoming data signals (and any data that they might be decoding and surreptitiously transmitting to Asna), our techies have broken their 'gramming.

They actually think they're blocking us with that crappy old stealth technology! What they don't know, is that I have a certain nameless friend who captured one of Asna's cyborgs and pretty much "teased" all the stealth 'gramming out of its brain.

Bepko da Silva deprogrammed it and deciphered the code, so while we didn't interfere with their work, we were receiving, recording and deciphering all the information that the drones were sending back to Asna. We didn't manage to capture his incoming data feed to them, but we did get the drone's outgoing transmission, as well as the data from the ROV itself.

I'd love to have found them trying to lie to us, but they basically verified what we already knew. The transmission's more or less drivel. It's in plain standard Terrax; however, it doesn't make any sense. It's random phrases, mixed with made up words, strings of numbers, oddly tweaked sounds.

I've got two of my brainiest MMs, and my double agent, working on the various phrases and nonsense.

Of course they're running it through millions of computer programs to see if they can get any kind of meaningful information. It's not the content that I worry about. The only thing that concerns me is that it is sending the transmissions bouncing off a satellite that can be traced directly back to Phaeron, Rehabco.

Now, we all know that Asna and his boys are the moguls of media on Terrax, but the MOB also controls the Retainage facilities on Terrax and Phaeron.

Originally, the MOB also owned the rights to the genetic

engineering of humanoids, the PHDs, so that they could program people to desire whatever products they had to sell. Whatever kind of 'gramming they had available, they created the viewership for. They imagined themselves brilliant.

Too bad they weren't really intelligent enough, far-seeing enough, to realize the negative implications of their conceptual programming ideas, were they to succeed.

We did.

When the Senselles first popped through the Time Rips, we started buying up huge chunks of shares in the MOB. We bought up the leftover genetic patents and obscure gen-mo rights that they weren't interested in. They were throw-away stock options regarding esoteric genetic engineering licensing laws, so they let them go for a steep price, figuring they had rooked the rookies. Little did Asna know, that was the beginning of the end for the MOB!

You know, that's something I never understood about the MOB. They had little or no technical expertise, how in the World's did they keep all this machinery functioning before we came along?

We had some clear advantages over them from the start. Our technical knowledge exceeded theirs, but it was more than that. We had the good fortune to know that physical objects existed within the strata of the earth itself, and that somewhere buried beneath the Hives and the Warrens, there were untold riches to be had.

We also saw the potential to harness more energy from the oceans, using existing OTEC plantships. Apparently their fear of the ocean had kept them from exploring the OceanDomes, and their tricky depths.

The MOB feared the oceans, they feared their environment. When we came, we knew that unless we could conquer and dominate the natural environment here (whatever it had changed into), that there was no way we were going to be able to survive here.

After acclimating to our new lives here, we eventually found ways to recycle raw materials. We financially supported space exploration, because we knew that unless we could get to other planets that we would soon run out of all the remaining raw materials here.

These materials are absolutely crucial in the manufacture of the spare parts and electromagnetic fiber optic cabling necessary for maintaining the communication systems that run the machinery of Terrax.

The general viewership doesn't concern themselves with such mundane matters as how they are supplied with power, food, water and air. Since everything has been provided for the PHDs ever since they can remember, they don't even think about where their power comes from.

The buildings themselves (the Hives) are living crystalline materials that collect energy – much like the photovoltaic or PV cells of Olde Earth.

The biggest problem with the Hives are that they require ongoing maintenance. Like any living organism, the biophotocells continue to grow in unruly masses, and tend to form sack-like growths that attach themselves to the outside of the buildings.

The growths (similar to the barnacles of Olde Earth's oceans) on the outside of the Hives must be scraped off by machine in order for the energy collecting materials to function properly. The Harvesting Machines are programmed to clean the outside surfaces of the Hives on a routinely scheduled basis, allowing the transient population of AGGIEs ample warning time to leave them.

We know the AGGIEs insist on domiciling in the Sadsacks rather than their WC assigned cubbies, so it's their responsibility not to get themselves "wiped" in the process.

Of the nine major Hives on Terrax, the most important Hive is the Melissae Hive, where the Senselles and the MagicoMathematicians live.

The techies, as well as a sprinkling of the elite BayBay viewership who have the best programming and editing skills, live in the other eight Hives, which are subservient to the One.

Elite Bay Bays are the talented few who can come up with the original 'gramming that'll keep the burgeoning mass-mind out there occupied, which is the only game in town.

The Warrens below the surface of Terrax are for the PHDs who've been devolved (sent DownTown) because of financial difficulties, the ones who can't earn enough chedits to support their vices.

Once devolved, amenities become scarce, with minimally sanitized water, expired protshots, barely re-purified air for their tiny sub-par cubicle, and only BASIC access to the Vidnet.

As you slowly languish from the rapidly increasing enviro-toxin exposure, you begin to consider death a welcome alternative. Occasionally, a desperate PHD will even call for the TDM (that's why the PHDs have fearfully dubbed it "The Death Machine").

They are disimaged and another drone rises to take their place.

Life goes on for the billions living in the Hives above.

The meshlings who can afford it can purchase the requisite "license to gen-mod level 3 tissue", and start up a nestling of their own. They begin their future generations in a test tube, where all life here begins.

When their nestlings come of age, they are put on a waiting list for the next available cubby UpTown, but if none are available they are housed in the Warrens for a short time. Whether your first living assignment is in the Warrens or the Hives depends on how much your procreators can pay to "place you".

The rest you have to earn, so it helps if you start creating valuable programming, right out of the tube.

My meandering thoughts are interrupted as a scatterstorm warning sounds, and I read out the data on the storm heading my way.

This scatterstorm is thick and thunderous, so I'm switching to manual controls. I love flying in weather like this, it's good for taking your mind off your petty troubles.

I know it's dangerous. Hell, it's practically the only rush left around – the only thing that really challenges me.

The bulk of my life seems to be repairing this machine or programming that computer. Then, the endlessly stupid gossiping, backstabbing and general mayhem that people cause; all the little people with all their little stories, the incredible drama of it all.

Man, I'm not interested in all that. I'm more of a hands-on kind of guy. I like to fix things, keep things going. Our most serious problem at present is our lack of raw materials and recyclable water, a problem that Vee is working on with the World Council through her Senselles. They're throwing everything they've got at it to get the World Council to release the funds for our next space mission.

We've got Arpad Reitan, the best H.O.R.S.E. around, ready to head out there and rope us in some water wealth. He knows the location of a rich comet field he says we can mine for tonnes of raw materials. He spotted one only 4.4 billion miles from Terrax, dubbed 2004DW, composed largely of metal hydrates, hydrogen and oxygen.

Water! Water is wealth. To get metal ore out of the mission as well, that's just gravy. Asna will be chewing on the inside of his lip when we finds out we've trumped his sorry ass.

Arpad discovered the field out in the Kuiper Belt, right outside Tonopah's orbit. He estimates it's over 800 miles in diameter, and I am slavering at the jaws to get a hold of it.

(Vee and her Senselles are also working on a little mystery that has cropped up. We need to know more about the genetic aberration that has been showing itself around, nicknamed "The Milliner" by the viewership. Who created it? Who turned it loose? When can we catch it? And, What the Jack do we do with it when we have caught it?)

Then, there's this other thing – the errant ROV, and the nonsensical connection to Rehabco, Phaeron. My take on it is that we have got to get out there to Phaeron and find out exactly what's going on at Rehabco.

The Retainage facility on Phaeron is completely run by maintenance servobots with basic programming. They aren't capable of making decisions, they have no autonomous power, and therefore they can't be bribed. Everything appears to be in order with the bots, from what we can tell from their data feeds.

The facilities are running at full capacity. None of the inmates are mysteriously dying or disappearing. The place is operating like a well run commune, yet I feel like there's something hidden going on.

Even though we've got cameras and microphones everywhere, and despite the fact that we're continuously testing the inmate population for emotional stability, it seems like smooth sailing, but in my gut I know that something's wrong.

It's way too idyllic – way too perfect out there.

Then we have this ROV, getting jammed up in the OceanDome doors, then frantically sending out that weird code. It looked like an innocent accident, but the ROV stopped precisely at the floodgate that controls the water intake from the ocean and sends it to the holding tank.

Once the ROV stopped there, it couldn't be reactivated. Then, right on cue the floodgate closed and clamped the ROV in its jaws, crushing it just enough to make it hell to get it out of there. While it's stuck there, it's allowing tons of water to gush in from the ocean (and that water has lethal concentrations of nitrous oxides, sulphur, mercury, lead and arsenic, along with hundreds of other carcinogens.)

It gushed past the holding tanks and starting to flood the lab areas. For the first time since the Southern OceanDome was built, they had to manually block off the holding tank overflow from the offices and admin areas.

That's when they called me. When the water started to back flood into the holding tank, they had to close that off as well, and they realized that the pressure would burst back into the building. The vacuum that was created allowed the water in, but it couldn't go back out.

It was as if some ghoul who knew how things worked, had rigged it so that the ROV would stop right where it would make maximum damage. I don't know if the ROV's spontaneous outbound transmission of data was part of the plan.

The whole mess reeks of Asna, but the confusing thing is that when he had his techies come down, he didn't send his main man, Barrage Rhyzin. He sent two Nisolle cyborgs down.

From our monitoring equipment, we gauged that they were transmitting exactly what we were receiving. Our stealth deciphering equipment confirmed that they were sending the exact same information privately as we were getting publicly.

So, whatever Asna says, it appears that he doesn't know about this, either. The fact that he was willing to send two of his own personal cyborgs down also indicates his interest has been piqued.

Although it could still be a sham, both of those facts put together indicate to me that this is something else. I am very anxious to fly down, get the raw data and examine that ROV for myself.

Here's to it.

As Wad Medani prepared to land his SpaceLite Shuttle, he got the strange sensation that after this, life would never be quite the same.

Distill the Pain

"why do i keep all my old diaries,
the years of pain and isolation
kept alive by the pen?

why do i choose to listen to those songs
that bring sad memories,
just to live them all again?

why do i wonder,
do i dream in the night,
if i had been different,

would it have changed the outcome,
would it have
been made right?"

- FROM DISTILL THE PAIN, BY THE BARD PHOENIX GAET

Yosaka- The Guardian

The Guardian, Yosaka, had just completed his rounds, checking all the security points as assigned by the MaxSEC Policies and Procedures Manual. This included a detailed check-off list of all the Retainees, the trustees, their guardians and, of course, the MOB attendants. He had e-skanned the perimeters, offices, entries and exits, making certain no untoward oscillations had occurred in the scanning equipment.

Once back at his post, Yosaka meticulously checked the power supplies-both incoming and outgoing- and the data trails, to make sure that no blips had popped up on his screens while he was gone the quick quarter-hour it had taken to walk the floor.

Making sure all his duties had been attended to, he turned to his percomp and called up his favorite hologram, an obscure old media gem known as a “home movie”.

Being an historian, Yosaka loved to watch the images of this sleepy little neighborhood dating from before the great WWWs of Olde Earth.

He gazed with scholarly fascination out the window of the comfortable suburban home, watching the plant life outside gently motion its leaves as if sadly waving goodbye.

He had memorized the names of some of the plants; over there were two potted palm trees, looking young and flexible, their long fronds bending to and fro in the light breeze; there were geraniums in a riot of reds ranging from ruby to pale pink arranged around the edges of the yard.

Toward the front of the yard stood a small, rather sad looking Norfolk pine tree that listed a bit to the left, its needles a dull green, indicating that this was the autumnal equinox, when even the evergreens that dotted the yard were still and sere. A huge jade plant stood at the southeast corner of the yard, marking that boundary, while a few feet away a bird of paradise bush stood with its leaves green and abundant.

One lone bird of paradise flower stood, its proud head sticking up on its thin neck, the fully erect plumage forming a bright orange and blue topknot. A hummingbird zipped from the geraniums, to the bird of paradise, and back again several times.

Yosaka gazed fondly at the moving tableau, remembering how he had purchased it at great expense and personal danger, from a disreputable looking fellow at the underground FLEE! market.

While gossiping with one of the other Guardians, a certain Malmo Dumjon, Yosaka had learned that certain exotic contraband items could, in fact, be purchased at the mysterious FLEE! market. The very name indicated the mindset of its sponsors, the "Fiesta of Losers and Egregious Err-ors!"

The door seal to the market could only be accessed by a secret revolving touch code. The catch, and there is always a catch, only the AGGIES could sell you the code. By purchasing this code you were probably, no make that definitely, contributing to their financial welfare. Although Yosaka had no desire to be associated with the Anarchists for an Above Ground Independent Egalitarian Society, there was nothing for it. He had to buy the Vee'd code from them!

Normally, Yosaka wasn't the type to get into adventures, but lately the sheer, unadulterated mind-numbing boredom of his job forced him out of his comfortable routine.

For the last five annuations, he had lived the exact same cycle over and over again. He'd go directly from his W.C. assigned cubby, via the public encapsulated slidewalk, to his night-cycle job guarding Retainees while they awaited their work assignments.

Then, after work, he'd travel the same route back to his cubby, his prayers and his study circle.

Being a rigid devotee of the Hopitu Shinumu Shaman, Esthana Thlehiya, he attended kiva on the full moons, the new moons, along with the requisite equinox and solstice ceremonies. He meditated daily and walked the Good Red Road in his journeys.

He was a good PHD, and he knew it. But during this last Wheel of the Year, his work-a-day world began to weigh him down like a pair of ferrocement boots.

Living comfortably within his 'gramming, he had never

before ventured out of his comfortable routine in 24 annuations, having simple tastes and being content with his class and work assignments. He earned enough chedits to fund his basic daily needs for what he liked to refer to as his three P's; upper grade plastiskins, protshots and pornies.

Of course, he had tons of social contacts, was peeped routinely throughout the day and night cycles, with full sensual input. Many of his friends who could afford to jack-in to the Sextel Network, would share a nanosecond of *Experience* with him; whether it be the flavor of a delicious meal, the touch of a lovers hand, the sound of incoming ocean waves lapping a sunlit shore, or the smell of roses on a warm summer's afternoon.

The name of the game was to feel something, anything; to cop a feel, to share sensations, to generate emotions, to VR together.

How curious, Yosaka reflected, that once freed of manual labor the human race had become addicted to emotions, the very thing they had given up "for their own good".

The average Jack didn't need to produce goods for sale, because the devolved BayBays who worked at the Big Four Industries produced all the manufactured goods on Terrax. The W.C., in their infinite wisdom, had assigned every viewer a free cubby to live in, free skinsuits (plastiskins of questionable quality), free low-grade (only semi-toxic) prot-shots and a continual saturation of cheap Eedia.

Without the need to work for shelter, clothing, food/water, or entertainment, there was literally no reason for PHDs to leave their cushy cubicle. They've been provided all their basic needs, danger-free. From that day on, the general population stayed holed up in their cubbies, sucking up electronic entertainment, and keeping the MOB fat and happy.

Most of the freecitizens (the viewership) of the planet "worked" by providing a few hours of editing services for the worldwide VideoInternet, casually known as the Vidnet, each day cycle or night cycle, depending on personal preference.

At least, the NAWALS ("the people who live in the Hives"),

had the luxury of never having to deal with the harsh environment of their world. The Enbees ("the people who live in the Warrens"), were another story. They lived in the Warrens, Downtown Terrax, and they were the muscles that pumped the life's blood of Terrax.

They were assigned to work in one of the Big Four factories; either they would assist in producing the skinsuits and protshots, or go to the other two companies, that serviced and maintained the huge machinery that ran the planet's water and air purification systems, and generated the power that kept the planet alive.

Yosaka was proud to have what he considered a real job. Every day-cycle he traversed the murky and dangerous underground world to get to the holding pens at Rehabco, Terrax, and back to his cubby in one piece.

It's true, he took the chance of suffering possible harm from environmental toxins, but being exacting with his personal security precautions, he had installed a sonic shower fixture that sprayed plastiskin on every square inch of his body. He donned a skinsuit over that.

Although the people DownTown were a rough bunch, he carried no weapons. Assault on a Guardian will immediately get you thrown OUT. No one in their right mind would suffer that, especially if you are unwilling to have the flesh melt off your bones in a matter of nanoseconds.

Besides, it was well known that Guardians are thoroughly trained in deadly hand-to-hand combat; Tai Kwan Do, karate, jujitsu, judo (as well as down and dirty street fighting tactics.) He feared no one.

The Guardians are the historians of Terrax, studying detailed information downloads from the TDM memory bank. They are privy to knowledge and ancient history that no one else knows.

They alone are the ultimate judges who decide the fate of the Retainees in their care. By observing their charges for an annuation, and comparing their behaviors and motivations with a profile of other criminals with similar behaviors, the Guardians can properly determine the appropriate work assignment for each Retainee. One that fits their crime.

It's modern day Justice, as meted out by the wisest among the PHD's, and no one questioned their choices. Rehabco, owned and operated by the MOB for the "edification of the safety of the viewership", is the closet thing to a jail this world has.

A Guardian's word is Law.

Assignments are doled out according to the seriousness of the offense. In an enclosed environment, one of the worst offenses is spitting volatile poisons into the air purification systems. This crime is rewarded with the lowest, dirtiest, meanest, and most dangerous work assignments available.

These miscreants will spend the rest of their lives cleaning the air filters (located on the OUTSIDE of the buildings) on the UpTown hives, or servicing the huge Harvesting machinery that clean the planetwide biophotocell arrays growing on the OUTSIDE of the OceanDomes. The Domes currently generate about 50 percent of the power that runs the planet.

The Last Great War transformed Terrax into a society consisting of four main class groups, with the usual smattering of lunatic fringe that accompanies all societies.

The NAWALS lived UpTown, their Hives dominating eight strategic points on the planet's surface. Housing the lucky few, the tens of millions of people who were allowed to live in AboveGround housing. These were the PHD's who kept themselves sealed in their environmentally contained buildings, in their completely safe cubicles, producing original electronic entertainment (Eedia) for the masses.

The rest of the twenty-six billion inhabitants of Terrax, the Enbees, are strictly consumers and editors of this massive glut of entertainment. The MOB discovered early on that continual entertainment is necessary to stave off the utter madness that having a life filled with nothing to do, can cause.

Entertainment is God here.

Until the Senselles came.

Besides the viewership, there are two monster power groups

vying for the wealth and well being of the planet, the MOB (Media Overlord Barons) who anticipated the shape of things to come and managed to own the proprietary rights to the old fiber optic transmission lines. They bagged the space station satellites, using them to spy on their competitors, the Senselles (this group includes Auntie Vee, her OTEC Panatechies, Wad Medani, and his MMs).

The Senselles are the only people on the planet who can sell an uncompromised sense of touch, daring to touch each other skin to skin, creating and selling the sensations on the most secured channel on the planet, the Sextel Network.

They offer a virtual reality jack-in that can transfer feelings and sensual data receptors directly to the viewer at 100% VR capability.

The exact details regarding the precise market shares held by these organizations are shadowy, well-kept secrets. Neither of these groups are publicly held. In fact, Oogling information on these companies leads to instant, although anonymous, punishment.

The punishment may be meted out in several not-so-subtle ways, like having the air flow to your cubby cut off, or suddenly you find your chedit limit drastically reduced.

You might even be falsely accused of a crime and sent to retention at Rehabco, Phaeron (off planet). From all accounts, the stay there isn't at all pleasant.

Turning his thoughts away from the sociopolitical to the purely entertaining, Yosaka wondered if he had enough chedits to jack-in to Damaria's Dance tonight. He wondered if she was "on" yet, and, desperate to have a peek at her Dance, tried to figure out a way to hack into his trustee's chedit accounts to siphon off the badly needed funds.

He simply had to find out what she was up to, not only because he was addicted to her hookah, but this morning he'd gotten a peep from B.R. instructing him to study Damaria De Rotiarte for possible weaknesses and infiltration points.

Yosaka was given explicit instructions to accomplish this

close-up study with his own funds, so there would be no links back to his MOB boss.

Yosaka thought of Damaria's pouty lips, and tried to concentrate on hammering in the security codes within Rehabco, Terrax. It took a few nanos, and then he realized that if he bounced the signal off-planet through a pocket of P-energy, from Terrax to the Rehabco, Phaeron satellite and back, it would appear that the Rehabco was requesting additional funds for their own trustee's accounts. Nothing suspicious about an additional fund-req.

He sweated it out, waiting to see if the Fail Safe Chedit Checker would fall for this.

It couldn't be this easy, could it?

He muttered softly to himself. Something was wrong, it was taking too long.

Starting to freak out, he almost pulled the plug on the fund-req.

Wait. The "Debit Approved" logo was dancing in ecstasy.

It was working!

Yosaka's MaxSEC clearances got him the extra chedits deposited directly into his account, no questions asked.

Suddenly, he couldn't wait another nano, and clumsily shoved the standard issue jack-in at his workcomp into his neck, sighing as Damaria De Rotiarte blossomed into his full sensory input.

* * *

Off-planet, the Rehabco Phaeron Retainees were instantly made aware of this rare opportunity to "ride the white lines" off of Phaeron's energy grid, and surreptitiously into Terrax' main energy grid.

They quickly gathered and banded their energy signatures into a communal group, in anticipation of making the jump.

* * *

Somewhere beyond the spiral arm of the World's Council planets, the Dark Ones amassed.

Waiting.

the moment

there,
for a moment, just a moment
my heart softened. and i wanted
to reach out and cry with you,
to hold you close again.
but the tear in your eye
was only a glimmer of light i saw there.
you changed the subject-
the moment passed.
like a cloud obscuring the sun
i felt a chill, then your silence returned.
and my heart hardened,
constricting my chest til i heard a crack,
like an ice floe breaking away
to float in the frozen ocean.
to wander the white vista
alone.

- FOR DAMARIA, BY THE BARD PHOENIX GAET

Damaria De Rotiarte took a slow draw off her garish, gold plated hookah pipe, and desultorily reflected on her life. As a Senselle, she had access to any place on the planet, she could come and go at will, whether it be UpTown or DownTown.

The price she paid was that she had to be available to interface for a least 60 percent of both the day and night cycles, a price she was willing to pay in order to acquire more sensual memory.

Sensual memory was the only coin of the realm, the only way to earn more chedits.

She fully intended to earn enough to buy a license for a Second Identity among the Sisterhood. Her primary identity was that of an Odissi temple dancer extraordinaire, exemplified by her pale white skin with its light smattering of golden freckles, set in a face framed by straight black waist-length hair, setting off icy blue eyes.

When she smiled, her canines showed just slightly in the sexy overbite which emphasized her pouty full red lips. Her dainty 5' 2" body was plush as opposed to plump, and when she danced all genders were instantly, hypnotically, inevitably drawn to her. Luckily she kept her demeanor as sweet and earthy as it was sexy, so the effect was unspoiled by the appearance of overweening ego.

Her patrician nose vented at tiny snort, as she thought of the newest shenanigans from the underground movement known as the AGGIEs. She'd heard that these raving lunatics actually believed it was safe to live AboveGround (meaning OUTside in the raw, completely toxic environment!)

Everyone knew such a thing was impossible. Period.

The last public execution (held centuries ago) involved the "live recording" of a WC Guardian shoving the poor offender out of an airlock. It only took a few seconds for his unprotected skin and bones and flesh to melt into a gray viscous puddle.

Whenever there was even a whisper of civil disobedience from the viewership, the WC issued, mile-high, public vidscreens would air the tape over and over. No more bad behavior.

Enough said.

Still, she was slightly intrigued to meet these rebellious PHDs.

The hookah had long since gone dead, its cold ashes providing no illumination for her. She wearily reached for her herb tin and reloaded the bowl. As the Day-Glo green and orange leaves caught fire, she inhaled deeply, closing her eyes and pondering once more what the AGGIEs intended to do if they *could* live OUTside. It certainly seemed like a rebellion with no justification at all.

She disdained waste, and all of this seemed like such an incredible waste. Just as she was coming on to the high she remembered to jack into the net, getting a double bang for her buck. She could earn chedits and get high, sharing her drugs with the world so to speak.

Laughing lightly, Damaria was in an extremely charitable mood, especially because she was dancing this night cycle.

Dancing is what she lived for.

* * *

Unfortunately for Yosaka, he was not to have the pleasure of watching her performance tonight, because a psyche-alert on his MOBster Retainee, Frank, had just sounded.

Damn! There would be no time for play this night-cycle. Yosaka checked out the vid monitors that accessed Frank's cubby.

The scene was one Yosaka had witnessed many times. Frank looked like he was having a mini-meltdown, which could easily be treated with drugs. Yosaka stared intently at Frank, who was sitting tensely on the edge of his cot in cell#78&615$79- Rehabco, Terrax.

Frank was slowly shaking his head from side to side, muttering under his breath.

Yosaka tweaked the audio up a mite to hear better, as Frank rambled on......

"it's this incessant, unending, almost-but-not-quite inaudible hum, that continues endlessly on and on is driving me bug-fuck crazy. Thank Vee, I've only got nine more days until my sentencing…" Frankie continued, unaware of the world around him.

It went on and on and on for some time. Yosaka kept the video on and turned down the audio, unable to listen to the senseless patter.

Yosaka always felt sorry for these thin, sad types. He'd seen 'em come in like this before. They'd go dingy from the constant noise that was a part of life underground, where Rehabco was built to encase society's miscreants. The constant noise, the lack of quality protshots, the withdrawals from their VRM addictions, could break the body and the mind.

By the time you were awarded your new work assignment, you were almost happy to go back to whatever comforts aboveground your new income could afford you. Let's face it, the system worked.

As an Historian and Keeper of Past Records, Yosaka could recite lessons he'd learned in training long ago. He knew the old stories by rote; how Auntie Vee and the MagicoMathematicians had changed the face of the ancient world.

As a balm to his bored mind, he let the information start to reel through his memory, while keeping an eye on his charge. Any significant changes in Frank's vital signs, and Yosaka would have the doc bots in that cubby in a nano.

* * *

It was known that . . .

After the World Water Wars of the 2040's, the Real Energy Crisis hit. As the enviro-economists had predicted, the supplies of natural resources were exhausted.

The governments who had subsidized the traditional coal and oil utility companies, in their greed to squeeze every dollar out of their oil holdings, hadn't properly prepared new energy sources.

Now, the world lacked oil, coal, and gas supplies, all the "traditional" resources that had been used to make electricity.

As a result, the fresh water supply had become irretrievably polluted, people were unable to drink it, or use it for any purpose whatsoever. The situation became critical, when Auntie Vee got the bright idea to encase the oceans in plastacrete, flash welded with bio-polycrystalline-microbial matter (BPMM) or biophotocells. The OceanDomes had been born. The Oceans were protected from additional pollution, and instead used to create electrical energy with ocean power.

The BPMM is a living organism that generates electricity and transfers it through the translucent plastacrete. You can still see right through the Dome to the boiling oceans below.

Although this left absolutely no open space available on earth, and no open bodies of flowing water, it did create electrical energy and provided the essentials for civilization to continue.

OTEC (Ocean Technology Energy Corporation) was the world's saving grace. OTEC was able to desalinate the ocean's water, providing fresh water, creating a mariculture industry that fed billions, and generating gigawatts of clean, renewable electricity, powering the whole planet.

Auntie Vee had patented the technology, convincing the failing and desperate World's Governing Council to harness the power of the ocean's waves with WEC (Wave Energy Conversion), a sustainable energy generation technology that has no ill affects on the environment.

The real beauty of the deal Auntie Vee negotiated with the government was that she ended up holding 75 percent of the stock in the company! This was a sweet deal considering the WC agreed to foot the bill for the ongoing maintenance of the machinery, thus making her the richest woman in the solar system.

The only downside of OTEC technology is the continually humming crescendo of incredible noise, the sound of the ocean fighting her containment. Sure, the plastacrete walls are several kilos thick and buffer the sound, but the closer you are to ground level, the louder it is. There is no escaping it, no naturally quiet place left anywhere on earth.

Add to that the general din created by all the machines running constantly cycle in and cycle out. We have moplanes (the personal motorized vehicles that are harnessed to the outside of a cubby, allowing you to jump into them and take off); the underground factories banging out their wares; the gentle susurration of the air filtration and prot-shot delivery systems; and the ever present buzz of electronic media going 24/7/365.

And so it is.

Yosaka turned his attention away from the ancient history of Terrax, and began monitoring his other Retainees, while listening to SNOOZE RADIO for the latest gossip.

It was only when Yosaka's attention was otherwise fully engaged that Frank allowed himself to stop the idiotic rambling and turned his thoughts inward once more.

He knew he wouldn't be closely monitored again for another 6 day cycles. He had to get his scattered thoughts together, because he had finally found his arch enemy. He had only 6 day-cycles to figure out how to kill The Milliner, and get out of this hell hole alive.

SNOOZE RADIO'S DAY-CYCLE CHATTER

Talking Head #1-

"Of course the AGGIEs are wrong, wrong, wrong as rust! No one lives above the ground, actually outside, outdoors, exposed to the elements! Every hatchling learns at first lesson that if you're exposed to Radzeeze, it will infect you and, you'll be dead within two weeks. It's a death sentence to be OUT."

Talking Head#2-

"It's true, but take a breather and clean up those flecks of spit from your faceplate TH1! We all know that most viewers reside inside the Hives or the Warrens, in buildings, but what I'm asking is.....what of the urban myths about the Heeshes, the PHDs who have been exposed, been thrown OUT, and lived, morphing from one gender to the other over time? Isn't it possible?"

TH#1-

"Of course not, you idiot! We've been over this a zillion times...."

* * *

Click.

It was no use. Even the local SNOOZE wouldn't allow Frank to escape his own dark thoughts.

You see, Frank is a Heeshe, one of the minute percentage of the population who have been exposed to the brutal OUTside elements, and by some unlucky stroke of fate, was rescued. "He" had survived. As he regrew his body parts, he had eventually morphed from one gender to other.

Frank/ie's story didn't begin with the unusual circumstances, she was originally hatched a woman. She got connected up to a mate and decided to pop the cork on a reprocap (a reproduction capsule) and have a nestling together.

At first, things were going O. K., then her meshling Teekum decided to "become an AGGIE." Since he had decided to become part of the underground movement, she didn't talk to him about it. Whatever illegal activities that entailed, well, Frankie didn't even want to know. Teekum viewed the whole thing as a huge joke, but Frankie didn't think it was one bit funny.

Apparently, one day Teekum was followed home from one of their meetings by a thug who waylaid him.

The thug jumped through the slightly open doorseal, and busted into their cubby killing Teekum, then their baby, right in front of Frankie. The monster then proceeded to methodically rape and beat her, finally flinging her ravaged body OUTSIDE. He not only wanted her to die, he wanted her to have the worst death possible.

But, she lived. She was dramatically rescued by a famous Senselle, Cami, and had survived, to become a living urban legend. She had morphed into a Heeshe, and was befriended by a mad old scientist, who was simply called Grandpa.

He'd become a good friend of hers, a mentor and father figure.

Frank/ie had made up her mind to take revenge on the murderous creature that had basically wrecked her life. He was known as "The Milliner". The problem was that, by the time Frankie had completely recovered from her rejuvenation process, The Milliner had already been reprimanded into Rehabco, Terrax, on some other minor offense.

The Rehabilitation Company wasn't intended to rehabilitate the offenders at all. It was a place to house criminals one Wheel-of-the-Year, until their Guardian sentenced them to their next work assignment level.

This is the punishment here on Terrax for any crimes committed. The security in a Rehabco is the tightest MaxSEC available on the planet. No one can get in or out without the highest security clearance.

Work assignment levels are "awarded" when you receive your holostamp. This invisible stamp is hologrammed under the skin of your right wrist, like a bar code. It reveals all your vital personal information; including your chedit limit and the type of work you are assigned. The more heinous the crime, the more hideous the work assignment/job you receive.

Once out of Auntie Vee's Uberlabs, Frank changed His identity, ditched his useless work assignment in one of the Big Four, and got himself a job as an MIS for the MOB.

Then he got himself committed into the prison system, once he found out that's where the Milliner was, in order to kill him. He didn't know how he would do it yet, but he had six day-cycles left to figure it out.

Rehabco, Terrax was always filled with Enbees. The Nowals never got punished for anything. When your sentence is awarded, you're free to go live and work wherever you can, as the stamp permits. Frank sat in his jail cell wondering how he could manipulate Yosaka into telling him exactly where this bad ass "Milliner" was being held.

Without intending for it to happen the whole scene of her mate and child-to-be being murdered, came back right before her mind's-eye.

It was like a nightmare that didn't go away, unless she happened to be morphing. For the first three weeks after morphing, a Heeshe doesn't remember who he/she is or anything about the past.

During one of those in-between times, Frank got on the wrong side of the MOB. Once Frank remembered that the MOB had ordered the hit on her/his family, (DAMMIT it's confusing) Frank went renegade and double crossed his MOB boss.

That's how Frank conveniently got himself into Rehabco.

Frank didn't remember much about his retrieval, but he knows he would be dead if it wasn't for Cami. Her Retrieval, which had been taped, had been replayed for Frank a hundred times, but his actual memory of it was a bit patchy.

Cami had rescued her, after Frankie had been tossed OUTside

by The Milliner.

Can you imagine? One of the biggest stars in the world happened to be the person who found Frankie!

Senselle is a play on words for sin sells and senses sell. Senselles bring awareness to people's senses. The Sextel Network even plays in the huge underground theaters that serve the poorest of the people, the BayBays who don't live in the UpTown high rise buildings AboveGround.

In Underground Theaters, you can experience any vice that you want. You recline on the lushly padded love seats, and hook a device into your neck port, or "jack-in" to a feelie. You are now able to experience whatever is happening to the actors on the screen. You feel everything that the person on screen is feeling (it's called 100% VR 'gramming.)

Cami has been a mega-star forever. She disappears from the public's view for a short while and then returns. When she comes back, her shows are wildly popular because she risks everything. Cami doesn't wear skinsuits, she has (physical!) sex with people, she'll even injure herself (on purpose) for dramatic effect! She's willing to experience every emotion, no matter how exotic, in an attempt to color this gray, rigid world.

I'm beginning to suspect that Cami has another role in all this power play, perhaps that of a rebel. Rebels don't try to change society, so much as to say "screw you" to it. They live how they want to live, without the structures imposed on them by society.

When I was pitched "OUT" and left for dead, Cami just happened to be "OUT" at the very same time. I imagine she could be a secret member of the Above Grounder's revolution or the "AGGIEs". AGGIEs believe in challenging the elements and not paying much attention to what society says is appropriate or inappropriate.

When Cami found me, she took a liking to me and rescued me. After my regeneration in Auntie Vee's Uberlabs, Cami introduced me to Grandpa, who would help me with the long recovery back to health.

Grampa, (the creator of the TDM) had been brainstorming a pilot program that would reverse the effects of the TDM. He wanted to create life rather than just preserve it.

He and a small group of MMs had found some ancient written materials regarding an entity called "the soul". Their findings supported their new-found theories on the re-constitution of living beings.

They felt justified in using their talents on me for this project. Their daily duties had, up to now, been to solve the world's most serious problems, and for once, this highly philosophical bunch decided that the life-preserving machine would be their next project. A veritable human think-tank, they believed it worth the effort and would brook no arguments about it.

The question they sought to answer: Why shouldn't we also preserve the lives of those whom society doesn't consider valuable? The TDM (Thermal Disimaging Machine), allows a person who is suffering from a terminal illness to voluntarily end their life. The thermal function of the machine dissolves the person's entire cellular structure, in a pain-free process. The disimaging function literally preserves or records every experience, memory and quality about this person within the machine, to save for posterity.

The TDM is conscious and has its own personality. It is aware of the person's reason for thermally disimaging themselves. It's not going to zap someone if they are pushed into the machine, it knows if you are prepared to die.

Of course, there is a huge debate about the TDM being used against someone, or whether it is evil. There are many different attitudes about it because there're many types of people.

The Enbees (more commonly called BayBays) call it "The Death Machine". They don't understand its higher philosophical value and see it as a threat.

The new project's purpose was to reprogram the TDM, to see if people could be reassembled.

Gramps was still in the prototype stages of the new machine, and yet somehow Cami knew all about it. She was undecided as to whether it was a good or bad thing, but she was instructed by Auntie Vee to take me to Gramps in hopes that he could help me survive with the mess they'd made of me.

I saw being with Gramps as an opportunity to heal, and then find a way to get revenge on the thing that haunts my dreams.

I knew I needed to get a job working for the MOB, then get caught by them doing some petty crime and get put into Rehabco, where he's awaiting his next work assignment.

Well, I'm here.

Now, I have to do it within the next six day-cycles.

For me, nothing else exists.

While I wait for my perfect opportunity, I'll distract myself by jacking-in to the next installment of Cami's Road Adventure "Mardi Gras, Done The Hard Way".

* * *

BENEDICT ARNOLD

The rest of the rides to LV were uneventful, except that as we traveled into the northern part of the state it began to snow ever so lightly. We called ahead to make sure Ben would be home, and that the coast was clear.

He seemed genuinely happy to hear from me, if a bit nervous. He was a medical student in his third year that I had met last summer while working for my grandfather. He said he was just nervous about an exam.

My grandfather's "Big Fun Curio Stand" was a regular fixture at the Indian pow-wows, rodeos and festivals in Arizona, New Mexico and Mexico. Every summer, my family would join him to help him sell curios, toys, hats, flags, dishes and magic tricks to the reservation Indians.

His friend, George, had a food concession stand, selling fair food like hot dogs, hamburgers and ice shaves. Last summer, George had hired Ben as an assistant, and in the long dull hours waiting between rounds of customers I would sneak back and forth between the stands to chat and flirt with him.

During that week together, I had become so attached to him, we exchanged addressees and promises to write each other. I thought I was in love, and every love song that came on the radio that summer reminded me of him.

I'd remember his tall, well built body, his sandy blond hair and green eyes. He laughed a lot and seemed easy going, his smile warmed me to my toes. I was flattered at all this attention from a man, especially since I was 15 and he was 21!!

He didn't even talk down to me or anything.

Anyways, once we sloshed into his pad, Cherry and I were soaked, not really being dressed for snow. We talked awhile and toweled off, and he asked some sly questions about the trip. He wanted to know if we smoked pot, and we lied and said "sure".

We trooped off to an old barn-like structure not far from the dorm room he shared with his roommate, and we smoked (our first!) doobie. It was a big thick joint of something he called "Colombian", in a purple paper with little glittery stars on it.

It was my first serious introduction to pot. Back home, pot hadn't infiltrated the high schools yet, and none of my runnin partners had access to it either. I was so nervous from the stress of the trip, I didn't know how tightened up I had been until I started to unwind a bit.

All the worries of our trip ahead and the problems we had narrowly escaped seemed to loom less large by the puff, and we were soon giggling like mad. Everything seemed funny!

At one point, Ben got hungry, he called it "getting the munchies", so we went back to his room to order a pizza.

By the time the pizza arrived, we were starved (the driver had gotten lost twice trying to find us.) He explained that these dorms were hard to find on this campus, you had to know exactly where to turn off.

Luckily, Ben's roommate was going to be gone the entire weekend, so after dinner Cherry took over the roommate's bedroom. Ben and I were left alone in his room.

Uh...Did I mention I was a virgin???

Ooooohhh yeah, a virgin, and nervous as hell and scared to admit it, even to myself. Hell, I had never seen a man's dick, limp or hard, but I didn't want to seem inexperienced or...uh... dorky.

I figured Ben, would, you know, do some of those romantic things a man does to get a woman ready. Get some juices flowing? Foreplay???

Apparently, he was in a hurry to get it over with, because he just started grabbing roughly at me. After a couple of uncomfortable squeezes to my boobs, he grabbed the crotch of my jeans and rubbed and squeezed some more.

I wondered if all this grabbing, (it was like he was kneading dough), was supposed to turn me on, because it didn't.

He hurried on.

I was simultaneously trying to push his tongue out of my mouth with my tongue, trying to draw a breath, and fumbling to tear off my pants before he poked a hole in them with his fingers. As he flung off his jeans, he stuck a blunt forefinger into my panties and banged my hips with his hips.

After more clumsy lunging about, he stuck his finger in his mouth, leering at me like I knew what was coming and plunged it into my twat without a gentle motion to it. I screamed, which he took for lewd encouragement, and slobbered on his other hand, then got himself slimy and hard in a couple of strokes.

I was disgusted as he used his finger, uncomfortably jabbing it in and out of me, but that was nothing compared to how it felt when he stuck his knee between legs to open me up, and rudely jabbed his dick into my twat.

He dove deep into me to the root of his hard, dry erection, and grabbed my butt to raise my hips for a better angle. Not glancing once into my face, he pushed on.

I was crying in pain, but his need was great as he banged away, putting his weight into it. I screamed "It---won't-----fit!"

Encouraged, he started doing push-ups into me, his dick slamming me harder and harder as my hymen valiantly tried to repel his efforts.

Finally, sweat dripping off his brow, he managed to pant out........ "you (slam) didn't (thud) tell (thunk) me (plunge) youwereavirgin!!!!"

This last bit was spoken all in a rush as I finally burst open, my blood finally lubricating his parched, irritated dick. He took two more strokes to rock me hard enough to get himself off.

Ouch! When he climbed off of me, I was vastly relieved. And sore.

I had never seen so much of my own blood, and looked around for something to mop it up with. I couldn't quite move yet. It was hard to tell what part of me hurt worse.

Suddenly, Ben excused himself, throwing me a towel. He said he'd been invited to a party upstairs, one for the house residents only.

Without so much as a casual "how was it for you?" he left. He didn't look back.

It was the last time I saw him.

The next thing I heard was muffled yelling, then a heavy pounding on the outer door. I saw the flimsy interior door to our room come crashing down, as it came flying off its hinges and fell to the floor.

I lunged into my clothes, getting dressed in barely enough time, when three uniformed policemen came storming in, guns drawn.

"You are both being detained for running away from home while still a minor. Get dressed and come with me now!" the first man roared.

We were both dragged, kicking, screaming and cussing, through the mushy snow in the dark of night.

We had been caught!!

* * *

Light through shuttered window

my soul is like the sun,
shining, illuminated,

too fiery bright to observe
directly.
that sun
hangs outside the window of my life,
fiercely gleaming through the partially-closed slats,
showing only
    one vibrant slice
        of one ray
            at a time.

By Kayla Raconteur

Arpad Reitan worked for one of the Big Four Industries, for the biggest of them all, OTEC. Arpad worked as a techie for OTEC's Underwater Repair and Maintenance, Oceanus Facility, Northern OceanDome.

He supervised the ocean retrieval systems, the automated ROVs (Remote Operating Vehicles) that traversed the ocean's depths, retrieving pieces of rusted machinery, metal, old aircraft and sea craft, anything that could be recycled and reused in the recycling plants of Terrax.

These are the raw materials necessary to make replacement parts for all the machinery needed to keep people here alive. Only these machines can clean the air and water, can maintain the filtration systems that cook up our protshots, and make our daily applications of Aerosuits. All raw materials to be processed has to come from the R&M division.

Arpad had lived in the Warrens all his life, which meant he was a DownTowner. To most any PHD, this means you are pretty much stuck in your allotted work assignment for the rest of your life.

Not Arpad. He aspired to be a space cowboy, formally known as a HORSE (Hydrogen Oxygen Retrieval Specialist, Elite). These are the PHDs who ship out into space for months at a time, retrieving comet-sized chunks of debris that contain metal ore fibers and water, a commodity that is in desperately short supply on Terrax.

He had never qualified for that work level because he couldn't meet the physical specifications. HORSE's had to be short enough to comfortably live in the space pods they must ride to "go roping".

It was his bad luck, ending up with less than well-off meshlings as his procreators. In their infinite...cheapness, his procreators had hired a 'grammer that was of dubious quality, and Arpad had ended up being a freakish 7'2" tall!

Being a "genetic mistake", he had automatically been consigned to the Warrens for the rest of his life.

Due to his extreme height (most PHDs are only 6'6"); he was originally assigned to OTEC's Biofouling Division. The Biofouling Division maintains the offshore wind farms by removing the naturally occurring growths (biofouling) that impede the turbines during their daily operations. He'd excelled in the leadership of the cleaning crews and moved up the ladder of responsibility within the OTEC organization.

After ten Wheels of the Year (he was a devotee of Esthana Thlehiya), he'd been transferred to the Underwater Maintenance Division at the Oceanus Facility, and was put in charge of maintaining the behemoth salvaging machines that trolled the ocean's bottom for treasure.

He'd also managed all the Aqua Boys, the underwater workers who repaired the huge salvage rigs, mentoring them while they worked in the most toxic environment on the planet. The work was physically demanding, not to mention dangerous as hell.

Arpad shouldered his responsibilities well. He didn't slouch one bit for all his 7'2" frame. He strode along, tall and proud. Thin, almost to the point of skinniness, he could be called lanky.

And he always wore a big huge grin, showing a full set of horse-sized teeth. He smiled like a lunatic, rain or shine. It made you wonder what the hell he was thinking.

He was very enthusiastic and spoke in loud bursts. Shy around most people, particularly gemmas, he enjoyed managing the servo bots and the Aqua Boys who worked on the rigs.

The salvage rigs were gigantic, their yellow Jack-in-the-Box head (where the computer's CPU is stored), sticking up above the waves in comical fashion. The rigs are 'grammed to trawl the ocean floor for kilos at a stretch, combing the bottom for scrap metal of any kind.

When they surface for repair, they bob about, looking like toys being tossed about the unruly seas.

Although he had seen them a trillion times, Arpad stared at the salvage rigs in fascination, eyes gone big and round. When a subject particularly interested him, he looked wide-awake in wonder.

The color of his eyes were an indescribable striking deep-blue. When you look directly into his eyes, you can see fine lines shot through the irises, like the striations in a clear crystal.

One eye had been damaged when he was a nestling, as he was taken out of his conceptionary tube. Upon unsealing the tube, there was some damage to the fetal tissue, causing one eye to be a slightly different color than the other.

One was blue, the other a light brown, which was rather disconcerting to look at. The pupil of the damaged left eye took up almost the entire iris or lens, which was also unsettling.

When he smiled, his eyebrows would arch over his eyes, and he would bat his long golden eyelashes. Then he'd flash that insane smile as if to ask, "What do you think of that?"

Heck, all you could do was agree with the guy. He was just so maniacally happy about every decision he made, that it infected everyone around him. He was completely optimistic, all the time. When he laughed, it was more of a giggle than a guffaw, a shy, girlish little giggle.

Combined with the fact that he wore his hair straight up off his head in a very short doo (about an inch and a half all the way around), he resembled a surprised gosling. One that had just popped out of its shell.

This morning he was thinking about his crew's upcoming day-cycle job assignments, when he overheard disturbing news among the techies. The latest gossip on the line is that some wily

PHD is gallivanting around completely "off-the-grid"!

No one had ID'd the individual's Universal Locator Code; there wasn't an official TDM data file of the person; and no chedits had ever been issued to them. Yet this "person" had supposedly killed a Retainee while still being held in Rehabco, Terrax!

It was an unbelievably bizarre story, but it was all over the net.

Dismissing the story as trash talk, Arpad continued to organize the daily work assignments, deciding where to station his crews, when a stray dark doubt crossed his mind about Dr. Bepko da Silva, the MM in charge of Arpad's work unit.

At OTEC, there was such a complex organizational structure that it was very hard to figure out what was going on and who was really in charge.

Even the doubts Arpad had right Now were being recorded into the TDM, the huge memory machine that records all the concentrated information, memories, thoughts, facts and experiences from all the PHDs who had ever died since the TDM had gone into effect back on Olde Earth, millions of annuations ago.

The physical TDM had survived the millennia and years of human abuse to the planet. The vast, intricate network of machines had grown into the doppelganger of a sentient being, although it wasn't capable of decision or independent action.

The lower buildings on the surface of the planet had grown into crystalline structures that made Terrax appear to be a fractured diamond, hanging in space. The older buildings were composed of a ReGen-polycrystalline microbe, capable of reproducing itself.

Galaxies of stars, having been disimaged by the Time Rips which had occurred in Terrax' ancient history, combined to lend only a weak energy to the planet. The thin sunlight of the stars fed the polycrystalline microbes, causing the hives to regenerate.

From space, Terrax resembled an exquisite multi-faceted gem, twinkling amidst the velvety darkness of space. She refracted every color of light in the visible and invisible spectrums. The crown jewel of the known universe.

Arpad vowed to see Terrax from space, one day. He felt sure that day of opportunity was quickly approaching. He had only to wait for the peep.

While he was waiting, he accessed his fav Senselle, Cami-Chameleon in her continuing series, "Mardi Gras- Done the Hard Way!"

* * *

As fear tactics, rats are vastly overrated!

So these cops take us to their little, dingy jail, quite unprepared for handling two wildcat teen-aged runaways, who were bent on shedding (more) blood. This facility wasn't supposed to house females at all!

Coming from routinely violent households, neither Cherry nor I expected to get out of a beating for running away, so there was no downside to trying to lie our way out of here. We stood silently, searching each other's eyes in the cold dawning light. We had the whole thing planned out beforehand and nodded our agreement to go ahead with it, now that we were caught.

Determined not to go back, we faced our jailers. I knew that my dad would go absolutely crazy when he caught up with me, so I had real good reasons for carrying this off.

I had devised "the story" while we were thumbing up here for just such an instance as this. We were going to act like spoiled, rich kids who had permission to hitchhike around by ourselves. We told them we were just bumming around and our parents were out of town, that they couldn't be reached by phone.

Cherry and I stood our ground, acting defiant and outraged that we were being manhandled this way. After asking us the same interminable questions over and over, both together and individually, they rudely threw us into a tiny cell by ourselves.

This jail house wasn't geared to accommodate women (or, more correctly, girls) so the employees were a little on edge. The main sheriff-guy had to be dragged from his bed on what should have been a cozy Sunday morning to have to deal with two runaways.

Sheriff Ball grimaced as he looked us over. One tiny hellion with black curly hair, flashing brown eyes and an attitude to match, and the tall, silent, surly one. What did she call herself, Cherry? He wished himself back into his bed.

The fact that Cherry and I were scantily clad, filthy, angry, foul-mouthed and loudly demanding our cigarettes, did nothing to improve the general atmosphere.

The main cop, Ball, a short, beefy looking man with hair that had gone to gray early, stared at us with the enthusiasm of a trout fisherman who had hauled up a pair of old boots.

"Who are you two and where are you supposed to be right now?" his eyes were very serious.

"I'm Camille May and this is Cherry Cooper, and boy did you guys make a huge mistake!!" I claimed with indignation. I took the initiative, hoping my surprise attack would place him off-guard. It semi-worked.

He partially raised one eyebrow, then caught himself and tamped it back down into place. His voice assumed its original hard nosed blankness.

"That's for the FBI to determine," he watched me carefully for signs of fear.

"My parents are mega-rich and when they find out you put us in JAIL", I raised my voice to an ear splitting volume and screamed. "It's gonna be your ass!!"

"That does it! I've heard enough of your foul mouth!" He slapped his desk hard, making his assistants jump.

"Officer, take these two to a cell and we'll see how they feel about this "story" of theirs in the morning."

He glared at me with loathing. Probably thinking about what he would do to his 15 year old daughter if she dared to talk to him that way.

I suppressed a shudder, imagining just that scenario being played out three days from now, when my dad would roll into town. I set my jaw, more than ever determined to convince them to let us out, tomorrow morning.

We were led to a small cell with one rickety bunk bed, each

steel frame topped with a filthy mattress. There was a toilet, toilet paper and a sink. No soap, no pillows or blankets, no sheets.

The back wall of the cell was no more than a grid of intersecting bars from floor to ceiling, with a cement walkway on the other side of the bars, to facilitate the guard's viewing of the prisoners. The guards would take turns walking the cells, checking to make sure each prisoner was still there, hour after hour. The entire cell was visible at all times.

Even the door to the cell had a little window in it at eye level, so one could view the prisoners, even when going to the bathroom. The guards made a little game of trying to catch us with our panties around our ankles.

After we decided whose bunk was whose- Cherry got the bottom one, I covered up the little window with improvised toilet paper balls. The toilet paper came in those individual squares of paper that are folded in half, so I got them wet, wadded them up into balls and stuck them all over the window.

The next guard to check the cell came over to the window, and after a moment of stunned silence, he swore mightily and walked all the way around to the cement walkway to make sure we hadn't escaped the locked steel door, or broken through the 4" steel bars.

As this extra little walk irritated our captor no end, I felt proud of my ingenuity. I figured if I was enough of a pain in the butt, maybe they'd let us go out of sheer frustration.

I fell into a limp little pile on my bunk, trying to pillow my head with my arms, and ignore the rancid smell coming from the mattress.

The Suits and The Story

I awoke the next morning with a start, not knowing quite where I was or why I felt so trashed. It was like coming out of a bad dream, into a waking nightmare. Then I remembered.

"Cherry, you awake yet?" She could be so quiet, it was hard to tell.

"Yeah" her sigh said it all. "I'm dying for a smoke tho".

Then it's about time to get these guards up off their asses.

"HEY GUARDDDDSS!!!" I yelled heartily.

The guard that morning was an actual human being, a trustee, not a cop, who took pity on our plight. He scoured the jail for extra ciggies, our first choice being Marlboro red, then Camels filtered, then Lucky Strikes, then anything except generic tobacco or menthols.

The trustee brought us the smokes and asked if there was anything else he could get us.

"The keys?", I smiled sweetly. I knew we had made an ally when he broke into a grin.

"No can do, but I'll see what I can do about a couple of blankets," he said. And off he went in search of supplies.

Cherry hadn't slept too good, she had gotten some bruises from being dragged into the police car. She also mentioned she had seen a rat on the floor last night. She was too tired to even yell.

"Let's go over our story one more time," I suggested, hoping they wouldn't catch any gross inconsistencies.

By the time breakfast had arrived (soggy, gray-lookin oatmeal, no toppings), we were fairly sure we could sweet talk our way to freedom.

We definitely hoped it would be before lunch!

We picked at the "food", smoked interminable cigarettes, then waited some more. We had not counted on the waiting part. The boredom, the little nagging doubts about the story, and about each other started to eat at me.

It was a very effective tool in breaking down our confidence, especially since we were champing at the bit to get out of there!

The day dragged on. When boredom hit critical levels, I decided we would play cards. This is not so easy as it sounds as we didn't have a deck handy.

I managed to cajole the guard on duty into giving me a pen, and then improvise by drawing the deck on the individual squares of toilet paper from the dispenser.

For the next few hours we played gin rummy and chain smoked. It took a loooooong time to shuffle the flimsy toilet paper cards, which suited my purposes just fine thank you.

Eventually, after “lunch”, they came and took Cherry away for questioning separately, then they came for me.

My guy was “a suit”, of indeterminate age (to me anyone over 30 is repellant and not to be trusted). I guess he thought he’d seen it all, but he wasn’t ready for me.

As soon as the guard left, I started my verbal assault, first on his intelligence, then on his competence, then on his chances for advancement in this one horse town. And I was just warming up.

“I hope you don’t like your job too well” I cooed venom at him. “Because when my parents find out where we’re being held, they’ll be furious”. I looked around the bare room with a snort of derision.

I looked him straight in the eyes and with a condescending little “tsk, tsk” type noise I continued.

“Do you really imagine someone would run away with no extra clothes, no money, nothing???” I asked incredulously. “God, how stupid.”

That was going a bit too far, I provoked him into showing me his hand.

“If you are from Nevada, how come we found an address book with addresses in PHH? And a map with points marked from there to here!” he was mumbling something about a smart-mouthed, spoiled little girl, fairly frothing.

Ooooooppppps. I had quite forgotten my rather damning address book, and the map. Oh well. Best to brazen it out.

“We were going to visit California, then we just changed our minds. Is it against the law to have a map?” I glared defiantly at him, and he sort of backed off.

“What about the address book?” he smirked; he knew he had me there.

“What address book?” It was a rhetorical question, and he knew it. I was playing for time to come up with a better come back.

I recited my prepared story one more time, even throwing in some desperate tears about how pissed my people would be to have to come get me.

"Can't you just forget this whole thing and let us go?" I asked hopefully.

Not all the crying is faked because the odds of me getting out of here before my dad shows up are beginning to look pretty grim. I start to think about how mad he'll be the next time I see him, and start crying in earnest.

The Suit had enough of my histrionics and returned me to my cell. Cherry had about the same luck with her suit. We smoked tensely in the darkening light of the evening.

It was nearly Sunday evening and we were mentally wrung out, having great reservations about the whole mess in general and physically ill from the stress, inedible food and cold nights on rock hard mattresses.

Our cards were long gone, having been used for their original purpose since the guards refused to re-supply us, due to the window incident.

We had run the entire jailhouse out of cigarettes, too. At last, after another round of yelling and badgering them, our favorite trustee, Sam, came up with a bag of Bull Durham tobacco and some rolling papers.

Even our good-natured Sam was fed up with us, told us it was the last of the tobacco in the whole place, so SHUT THE HELL UP.

We subsided. Resigning ourselves to another long cold night of boredom, we went over our individual interviews one more time for clues as to our status as prisoners. It appeared they were playing it safe, holding us there until our parents arrived. Apparently they had been called on the first night we were taken.

Cherry had managed to eke a bit of information from her guard, who thought of us as defenseless little girls. He outlined our case for us, saying that it was obvious from the address book and the map that we were, in fact, runaways.

Turns out, my father had second guessed our destination and alerted the town police long before we arrived. The local police had contacted my dear beloved Ben, who was told that if he harbored us and was caught, he'd go to prison for harboring runaway minors.

(So he got his nut off, then turned me in to the pigs with his cum still between my legs!!!!)

The guard also let on that the Sheriff gave us some credit, remarking that he had almost been convinced by our bravissimo performances that it was a case of mistaken identity. Except for the fact that we didn't know diddly about Nevada- where we claimed to live.

So close.

The guard was sympathetic to our plight, and Cherry almost twigged to the truth, but remained obstinately silent.

Just then, I began to notice the voices in the cell next to ours rise in agitation. The inmates were speaking Spanish, which I understand only a little. Despite my ignorance of Spanish, I'm getting scared because the nature of the argument they were having with Sam, the trustee, was becoming plainer and plainer.

It all boiled down to trouble for us, as the inmates tried to convince Sam to open up their cell, then our cell, and then turn his back as they came in to rape us.

They were very persuasive, first wheedling Sam about his questionable manhood, then offering him seconds after they finished with us.

I listened, concentrating fiercely to try and understand the way things were going. Cherry, alert to some sort of problem, sat up and opened her hand as if to ask " What's going on?"

I explained it to her, and since I wasn't sure how the argument was going, I suggested she get on the top bunk with me and prepare herself to fight.

We quickly decided that if they rushed the door, we were going to jump on them from the top bunk, and try to break their necks as we choked them.

I was not going down without a fight. We sat perched on

the edge of the bunk for what seemed like hours. My heart was beating painfully, and I couldn't stop from holding my breath.

Finally Sam came up with an idea that shut them up. He had offered to turn his TV toward their cell and let them pick a show, if they'd just give up their argument.

Thank God for TV!

It took a couple of hours to get my heart rate back down to normal, when I fell into a fitful sleep.

* * *

After sharing the rousing adventure with "the girls", Arpad felt charged up with energy. He needed to calm down in order to make his structural presentation to the OTEC big wigs.

They wanted to know how much damage that errant ROV had done to the Southern OceanDome, and how much enviro-leakage had occurred.

To this end, he played a couple rounds of Quadra dimensional chess with three other roomies from his work unit. Around the Big Four, the Quadra leagues are grouped into houses, and we play four against four.

Which makes it perfect. You would think that the games would end up being tied, being two against two, but it never ends up that way. The odds are always three to one.

Always.

I don't know why, but the number three is significant to the powers that be (Auntie Vee).

After his Quad-DIM wins, it was easier to put together his crew's work schedules. Their primary mission this day-cycle is to contain the toxic seepage and enviro-contamination to the Warrens located directly beneath the Dome damage.

The Southern OceanDome encased the raging oceans composed of mercury, perchlorate, lead, petroleum byproducts, industrial wastes, PCB's and heavy metals. Although still called "ocean water", it's more like battery acid than water. The weird chemical soup combines to generate electricity, and when the electricity is retained for storage, so much the better. The problem is, the stuff eats up almost anything in its wake, making it (almost) impossible to keep the machinery going that works in contact with it.

To make matters more complicated, Bepko da Silva and Koutsakos, another MM, are heading up a research project for another structural problem, one that has nothing to do with the ROV's damage. Apparently, in the construction of the Warrens (located below Terrax' surface), they ignored the instability of the earth surrounding the OceanDome's perimeter.

The Hives above, having grown larger and denser than originally anticipated, are compacting down the earth located directly above the Warrens. We lose about one Warren Level every six annuations, and the Enbees are complaining of genocide by the UpTowners.

Can't say as I blame the naysayers, blaming the WC. The crushing, growing weight of the Hives above will eventually demolish the Warrens, and all the BayBays with them.

The class war makes no sense to me. I see the necessity in all beings. After all, since the Nawals produce entertainment streams of media for the viewership, who will consume the product if the BayBays are eliminated?

Who'll buy their edited 'gramming for the Vidnet (that the MOB controls and the Senselles dominate)? The resounding silence would deafen the world!

At any rate, I'm not really into politics. My mind is preoccupied by the news I just heard about a new medical procedure that could possibly start shortening my overly long legs. In case that utterly fails, I'm working on the other end of the problem with some OTEC hardware techies.

We're in the process of building me a newer, larger space vehicle that can accommodate me just the way I am. I'm gonna go out in my SCOW, grab chunks of hydrogen and oxygen in the form of small planetoids, and come back to Terrax a hero!

I'll cart back not only water but power for the biophotocells on the planet. This is my only way out of being a nobody, of beating the tedium of being a DownTowner Techie forever.

With all these distractions, I think I'd better spray on a second and third coating of Aerosuit's A-PLUS Aeroskin. I'm being particularly fastidious because today I'll be working with that battery-acid like fluid down below, and I've seen what it can do.

Machinery made of a hybridized laminated ferrocement and reconstituted titanium steel alloy (one of the strongest materials known to PHDs), can disintegrate into pudding within seconds of contact with the "water".

Arpad was careful to keep his personal level of enviro-contamination to a minimum, now that he had a reason to live. He was going to get revenge on that back- stabber, Bepko da Silva, and the only way to get access to somebody on that level was to be on that level. And a HORSE could spit an MM in the face, because they were considered roughly equal in status.

Arpad smiled his maniacal grin, winked at the reflection of himself in his vidscreen and made his way down the CAT (Connector Advancement Travel) chutes and into his waiting moplane.

Oh yeah. He was a pilot. And no one was going to keep him from flying his own craft.

* * *

Murder Most Foul

Frank/ie woke up this morning knowing it was the perfect day for murder most foul. She could feel "the transformation" coming on in her bones, and knew it wouldn't be long before her body morphed from male back to female, and the gig would be up.

She knew it was coming on because she thought of herself as Frankie this morning, as opposed to Frank. It was always this way right before the change. She would begin to have memories of her old self, her old body, her old life.

It was so different then.

But she couldn't allow herself to think about that right now.

She had to concentrate on the plan that she had hammered out, on how she would kill "The Milliner" today.

* * *

I school myself to complete calmness, as I have learned by watching virulcasts from Esthana Thlehiya's sacred Kiva rituals. I access my Nakaiya aural tapes, and, as I hear the first thin strains of the eagle's cry, prepare to inhale the Breath of Fire, Pranayama.

Gathering and focusing my energy while sitting in a comfortable, cross-legged position, my legs are pointed back in towards my body. My field of energy is thusly contained, as I build up my inner chi.

I raise my arms up, my elbows approximately level to my shoulder, my forearms at a 45° angle, palms held upward and cupped in supplication and celebration of the energy I gather in honor of the Goddess of Life.

Starting with my pranic deep breathing exercises, I allow my belly to remain soft and full. I inhale until my belly extends. I hold it for three seconds, slowly exhaling until all the air, until the last bit of breath has gone out of my lungs.

I repeat the deep cleaning breath three times (any more than three and it's simply hyperventilating). On the last exhalation, I begin the last step in the Breath of Fire, as I inhale until my belly extends, then quickly inhale and exhale ten sharp, shallow breaths. It is critical that absolutely all the air is traveling in and out of my lungs on every exhalation.

This exercise cleans out your lungs, because you're getting to the very bottom of your lungs this way, increasing your energy level by increasing the clean blood flow throughout your entire body.

After the sharp in and out breathing, I windmill my arms around, scooping up more energy. I condense the energy into a visible ball in front of me, holding my hands opposite each other, about a foot apart, I cup my fingers, feeling the intensity and life within that sphere of pure energy.

Playing with the ball of energy, I can feel it grow, feel it oppose itself, feel the energy I have focused to do with what I will (as long as I intend to Do No Harm.)

As I feel the energy moving up my Chakras, from the base of my spine, to the top of my head, I bless the energy and gather up enough to do the task at hand. After raising the energy, directing and guiding it to purpose, I release it.

This is Pranayama, the Breath of Fire.

* * *

As my spirit journey ends, I return from my trance state, in better control of my body. Realizing that every action, every movement I make today must appear perfectly normal, I review my plan to be sure that not one detail is out of place.

Later, the Guardians reviewing the daily tapes mustn't see a single action from me that will bring me unwanted attention.

As a trustee for the UVPs (Ultra Violent Programees), part of my job is to replace or repair all the broken light fixtures for the UVP Ward. I prepare for my work shift today as I have done every morning for the last three quarter turns of the Wheel.

I step into my sonic shower and stand here, arms akimbo, as the machine lightly dusts my entire body. I put on the Trustee uniform consisting of the garish purple, satin pimp-pants; a brilliant white, Loewi long-sleeved blouse with ruffles at the wrists and lapels, and the matching purple coat with tails. This style is all the rage with the Trustees down at the UVP Ward, and although I loathe the color scheme, I don't want to look out of place.

(Jack it! Even as a woman I didn't dress this foppish!)

Adding several strands of faux-gold chains, I top it all off with a soft, floppy hat encasing a secret compartment that holds the all important nano-needle filled with "O-Zone".

Since entering Rehabco, replacing the light fixtures had been at the heart of my plans to kill The Milliner. I had been the one to convince the Guardians (and their MOBster overseers), that the lighting fixtures caused the Retainees to become more violent.

Gramps helped me acquire the falsified "official W.C. authorized studies" necessary to convince the Rehabco officials to change their current fiber-optics fixtures to a soft bulb. The studies claimed that the light from these new bulbs had a significant calming effect on the Retainees.

That's hokum. The truth is that the soft bulbs can be pierced with the finest cryosurgery needles, thereby being injected with a deadly odorless, colorless poison.

Luckily, the MedCo Dispensary carries plenty of the specialized needles that I'll need. They use them to prep the Incorrigibly Violent Offenders for cryonic storage, before being shipped off-planet to Rehabco, Phaeron.

Whistling a tuneless song, as I always do when beginning my work rounds, I looked directly up into the monitors and smiled that harmless little grin the Guardians know so well.

I know I'm a particular pet of Yosaka's, but I'm going to have to watch out for Yosaka's debating opponent, and my nemesis, Malmo Dumjon.

Malmo the Guardian may be cynical, but he isn't stupid, and he has a very good nose for sniffing out the truth. He's been watching me like a hawk ever since I've been inducted into Retainage three quarters ago.

Malmo, while believing in the necessity for violence in PHDs, doesn't condone any behavior occurring outside of one's GCR (Gen-ennge Character Registry.)

To hell with Malmo.

He hasn't seen his family torn apart, mutilated in front of his very eyes, he's never been beaten and raped, then tossed OUT to die. He hasn't been turned into a freak without a future.

In Malmo's own words, "feckit!"

I know I'll be the first one Malmo will suspect when they find things amiss, come night-cycle. That is why I have to control my every breath, and never miss a beat.

It's going to be a long workday, I realize as I look down the interminable row of cells, cubicle after cubicle, for a good quarter kilo in every direction. I begin by going to the first light fixture in the first cell. I set up my bright yellow Little Giantess Stepladder, using both hands to climb the ladder, careful to watch my feet as I take every step up the ladder (appearing to be a bit clumsy figures into the plan.)

Taking a filament between my teeth, I reach up to access the hatch for the fiber-optic lighting cables with one hand, and with the other, I clutch the ladder as if needing support.

I carefully remove the old filament, placing it in my left pocket protector and replacing it with a new titanium/platinum alloy filament from my right pocket.

Continuing to whistle, I switch from Japrap to an O.E. standard by Roger Miller, "King of the Road." As usual, during this part in the song, I wonder what in hell a stow-ghee is?

In all my little chats with Yosaka, he has never been able to come up with the answer to that question. I guess it will remain one of those mysteries of time.

I make sure to move slowly and consistently, making the sequence of my movements precise, almost robotic in their repetition. This is a dynamic NLP technique employed specifically for the benefit of those who are monitoring my work today.

Soon their minds will mentally "tape" the event, and as they watch me, they'll mentally see it replayed over and over again in exactly the same way. Eventually their minds will replace the actions they are watching with that mental tape, in order to protect themselves from the mental boredom of having to watch me replace 1000 filaments individually.

It is a trick of the mind and the eye, a trick I am willing to gamble my life on.

I've researched how the mind works, having nothing but time on my hands during my long, long "recuperation" in Auntie Vee's Uberlabs. I yearned to know more about why a person is violent, and how to control the tendency to violence in some people.

In the course of my studies, I discovered that genetics predisposes a person to violence, and so cannot be avoided entirely. I also learned that you cannot control the violence in people, the individuals will either express it openly, or repress it until it is exhibited in some fashion, whether actively or passively.

Curiously, even with the connections I have through Gramps, the Senselles and a couple of MM friends that I've made, the available research came to an abrupt halt. I ran into security firewalls that made Max SEC look like kinny-garden.

I came to the conclusion that the MOB's gen-ennges had something to do with controlling the predisposition to violence among the massive viewership of Terrax. I think that they are concentrating all of the violent genes within a few of the populace, allowing violence to occur naturally, while keeping it out of the strains of the working classes.

Neat trick, if you can pull it off. It makes sense in a sick, egoistic, controlling sort of way.

By Vee, I don't care. They can have their nasty little secrets, as long as they afford me mine.

Because I'm the real thing, a living breathing urban legend; the proverbial monster under the bed come to life and knocking on your door. My existence disproves the MM's little theories about keeping people indoors, whether aboveground or underground.

But I'm no danger to them, believe me. I don't want to change society, instruct society, educate or save society. Leave the willing viewership each to their own vices.

Hell, I can't even justify my own pathetic little existence. All I really want to do is kill the savage unclean beast that destroyed my life, and gave me a gut full of emptiness in return.

DAYYYUUUUMMMMM!

I have to stop thinking about the past. I just hesitated in my whistling, stopped and started up again.

I take a mini Breath of Fire to calm and center myself.

That's better.

I am now at that the 600th cell, and this is the home stretch. My research on human psychology exposed an interesting fact, it claimed that when two-thirds of the task is finished, the mind jumps to the next task. Banking rather heavily on this, I had scheduled my work route so that the Milliner's cell#666 would be just past the 2/3 point.

I figure this way my overseers will not catch the fraction of a nano that I will deviate from my routine.

Acting as though I had done this all my life, I set up my bright yellow Little Giantess stepladder, using both hands to climb the ladder, careful to watch my feet as I take every step up the ladder.

Taking a filament between my teeth, I reach up to access the hatch for the fiber-optic lighting cables with one hand, and with the other, I clutch the ladder as if needing support.

I carefully remove the old filament, placing it in my left pocket protector and replacing it with a new titanium/platinum alloy filament from my right pocket.

Descending the ladder for the 666th time, I look down as if I'm unbalanced for a nano, as I quickly reach with my right hand (unseen by the cameras) and pluck the poisoned filled syringe out of its hidey-hole in my visor.

As I close the access hatch, I plunge the needle into the filament, discharging the undetectable, heat-activated gas into the light fixture.

I then continue on down the line, replacing every filament in the rest of the UVP Ward exactly as I've done to the first. I know that during the Milliner's evening exercise routine, he will keep the light fixture on for the additional amount of time necessary to create a chemical reaction between the filament and gas.

Sometime during the night cycle, quite without warning, the gas will discharge out of the fixture as it turns into a fine poisonous mist that will settle into his lungs as he sleeps. This chemical reaction will cause his inner organs to melt, spreading out to all his tissues from the interior to the exterior.

Unfortunately, it will all happen too quickly for him to suffer much. Fortunately, it will be too quick for him to call out for help. By the time the cleaning bots come around next day-cycle, they will assess the mess for what it is, human waste, and suck it up like all the rest of the filth on the floors.

* * *

The next morning cycle, I awake hearing the Ward abuzz with the news....someone is missing...someone has gotten out of MaxSEC! At this time of cycle, I usually tune-in to Esthana Thlehiya's Sunrise Circle, and nothing must seem out of place, so I do so today.

But as I tune in, I can barely stand not knowing what the Vidnet is saying about all this. Will I be found out? What will they do to me?

NO! I mustn't think this way. If I concentrate on the Circle, I will forget. Staring at the monitor as though it will save my life, I begin to lose myself in the ritual already in progress.

* * *

This is SUNRISE CIRCLE Programming from Esthana Thlehiya, the Hopitu Shinumu Shaman, brought to you by: The Sextel Network, "Where The Customer Always Comes First!"

I woke up this morning and I felt my motivation dragging a bit. I got up and noted that in only a fortnight the winter solstice will be upon us. I can feel a cold snap in the air this morning as I roll out of bed.

Quietly entering the Kiva (my sacred prayer chamber), I honor first Kockyang Wuuti, Grandmother Spider of the East with a humble bow. I turn then to the South, naming Tawa, our sun god, our god of warmth and creativity and illumination and his sister goddess, Pele, who is his equal. She brings the creativity and sexual passion of fire, as well as the warmth of compassion, of friends and family. I bow deeply.

I turn to the West and hail Palolokon, the great god of earthquakes, water, the sea, and rain. He is the one who brings rain clouds that wash down the sky and give us clean water to drink, for our crops and our animals. We need him to gift us his water that our corn may grow, for our pumpkin, our squash, and our beans to sprout. I take an extra moment to say a little prayer for the absence of snow, at least until I've completed my journey.

Turning to the North and Huruing Wuuti, She who is Corn Woman, Earth Woman, Earth Mother. She is the Goddess who brings all abundance on Olde Earth. She is responsible for the crops and for the early spring, for the full late harvest, that will survive us through the winter when there's nothing growing and not much to hunt.

Completing the circle, I return and hail Kockyang Wuuti, the Creatrix of life itself. She spins the web of life, creating our stories from beginning to end. Ancient in southwest native tradition, we acknowledge and thank the four quarters of the Wheel of the Year, along with Sky above, Earth below, What is Without and What is Within.

The Hopitu Shinumu see that the world is filled with life, and that all things are in flux and balance, each to their own cycle. We perceive humans as natural beings, our souls commingling with other living creatures being a part of All That Is. Humans are not below or above other creatures, but as life is within all creatures, as the Gods themselves are within everything, as the Goddess herself animates everything, so we are God and Goddess. We are life itself.

This morning I prepare myself for my Soyal spirit journey, so that I may know what I will teach The People on Soyal, the winter solstice. Winter solstice is a time when we welcome the Kachinas, our teacher spirits, to live among us for the first six months of the year, from Soyal to Niman, the summer solstice.

The Kachinas are helper spirits who have come to us to give us messages from our revered ancestors, the Blessed Star People. They teach us how to walk the Good Red Road, they remind us of those Spirits we have loved, who have gone on before us, be they animals, plants or humans. During this time in the Wheel of the Year, we reconnect with all the beautiful beings we've ever known, whom we no longer have contact with.

There are hundreds and hundreds of Kachinas, each one different, because they are the spiritual embodiment of the "medicine" or lessons of each living creature.

For example, if you see the domesticated felines that run

around the camp, you know that most of their time is spent playing, eating, and sleeping, in between their many naps.

What good are they, you may ask? What medicine or lessons could these lazy beasts impart?

Well, my friend, if you have ever been seriously injured, and there is great need for you to be calm and lay still for hours on end in order to properly recuperate, you will need the medicine of Cat.

If you haven't found the secret of cat medicine, you'll find that, for the first time in your adult life, you can think of doing nothing but returning to those daily activities that you were once able to do.

But, in your current state of physical disability, you cannot do much more than sleep, eat, perhaps have a brief visit from a dear friend, and then fall right back to sleep.

When in this condition, if you are still compelled to continue to accomplish those daily activities, you will not heal.

You will not heal, unless you learn the medicine of the Cat, the medicine of feline, the medicine of the ancient Egyptian Goddess, Bast.

Bast will teach you the value of lolling about all day, doing nothing but letting nature take its course, and feeling completely comfortable doing so. That is the medicine of Cat. A gift from the healing Kachinas.

Another important daily medicine to appreciate is the medicine of Dog. The dog is loyal. The dog lives in a pack. The dog shares responsibility for the raising of the young. Both male and female dogs help with overseeing the pups, protecting them and teaching them how to hunt.

The young are there to play, and the elders are there to take care of them, so that the adults can do the work of daily life. That is what the medicine of the dog is, to teach us that we are a tribe, that loyalty is more than a survival mechanism, that honoring each member of the pack helps to grow and enrich the lives of all.

There are many medicines in the animals all around us. And

these medicines will come to you, if you will observe the natural cycles of life all around you. To observe is not to merely see. It is to see with meaning. There's a reason why a particular medicine or animal comes into your life.

Today, I will go deep into my own subconscious spirit to reconnect with the medicine of Deity itself. To reconnect with lessons that I must bring back to my people, so that I may teach them what they need to know to survive, but more than that, to love life.

I want them to understand that Life is a Wheel, there are patterns that repeat, that go 'round and in time, will return. Whether you are within sorrow or joy, that time will pass, and there will be a time again for joy or for sorrow.

When I say joy, I mean to enjoy the things you have while you have them. Don't simply count up the stuff you have. The Wheel of the Year will turn, the Wheel of Life will turn, and the next day your Life may not be so abundant. Still, there is peace in that knowledge.

In preparation for my journey, I've come into my sacred place, the Kiva. I have set up the boundaries of the circle and set the intention. I set the purpose to go within and bring back the lessons that I will need to teach at the next full moon.

My subconscious mind lives in the West, the water's depths and currents run strong in my veins. Facing West, I begin The Tree of Life, an ancient energy focusing exercise that will connect me with the phantom energy, the dark energy in the universe. I reach for All There Is, and All That We Cannot See.

Holding my arms at shoulder height, straight out, palms up, I imagine my legs being connected to the earth by a silvery rope. It's glittery, like sparkling chains of fine silver, connected from the middle of my being, straight down my spine and into the dark fecund earth below, down through the floor beneath my feet, through the levels of ferrocement, the buildings, foundations, and ground, shooting directly into the soil deep beneath the surface of the planet.

I can palpably feel it.

I am its strength.

* * *

Frank snapped his attention back to the Now, as the Rehabco public access system began virulcasting Cami's Grande Finale to her infamous adventure, "Mardi Gras- Done the Hard Way." (Brought to you by the Sextel Network.)

The Retainees had been talking about its release for weeks. The show promised to have a knock-out RQ. Frank had even placed a bet that Cami would be seriously injured during the finale, so it would be out of character for him to be disinterested.

Frank dutifully turned his outward attention to the Vidscreen in his cell, although his mind was a trillion miles away.

Cami's voice crackled as it came roaring over the scratchy, antiquated VidNet Broadcasting Band (the VBB).

"Mardi Gras- Done the Hard Way." ***(Brought to you by the Sextel Network.)***

"OK, we've definitely been caught. We're not going to be able to talk our way out of it this time," Cami thought to herself.

At least I almost had the head honcho of the jail believing our story.

Ok, maybe not in reality. But give me some credit for being believable, he was looking a little concerned when I didn't drop my story.

I kept insisting that, in fact, I was a rich babe hitchin' around with my girlfriend for summer kicks while my folks were out of town. For a moment, I almost had him believin' that my parents were about to return any minute, mad as hell, ready to lambaste them all.

Get them fired. Make them pay.

Too bad he didn't go for it and let us go. No such luck.

Since Sheriff Ball had already contacted our parents, it was too late for him to change his mind. He was committed to his choice, although that didn't make it any easier for him to live with it.

No matter how convincing I am, he'd found physical evidence indicating we came from California.

Damn.

Turns out, my dad and Cherry's mom are on their way, driving three days straight to come get us.

You can just imagine the mood he'll be in when he gets here.

Or, maybe you can't.

Let me fill you in on some of the juicy details…

First of all, my father lifts weights; he calls it "pumping iron." He has immense upper body strength. Muscles.

And, he has total control of my family.

What he doesn't have, is patience. Or control of his emotional reactions. Or respect for women.

Unfortunately for the women in our family, his mother had been an angry woman, was bitter at being stuck raising children alone. She punished my dad for being male.

He, in turn, punished my mother and sister and I for being female. And around and around it goes.

He dislikes people smoking around him because he has sensitivities to every smell in the world. We aren't allowed to burn incense in the house, or to wear nail polish or perfume of any kind.

He controls our family with an iron fist.

He may dislike women, and all their affectations, but what he HATED, was women who wore make-up and nail polish, and talked a lot, and smoked.

And that is exactly what Cherry's mom did for the entire three days, non-stop, all the way from southern California to northern New Mexico.

He was livid when he strutted through that jailhouse door.

I mean, here I am, sitting directly across from Sheriff Ball. As soon as my dad stormed in the Sheriff stood up and extended his hand, as if to introduce himself.

I saw my dad in my peripheral vision, but didn't even bother to glance at him.

As soon as he saw me sitting there calmly ignoring him, his face started turning red, the veins popping out of his forehead. He sputtered to get words out.

I looked over at him very slowly, and then turned back to Sheriff Ball. I unflinchingly met his eyes.

"I have never seen this man before in my life," I purposefully stated.

For one precious second (I'll savor this moment forever), Sheriff Ball's eyes went slightly blank; his mouth sagged. His jaw didn't exactly hit the floor, but I really caught him off guard with that one.

I stared evenly into Sheriff Ball's eyes and saw a moment of panic dawning there as he considered that perhaps he had made a mistake.

Had not my dad leapt the two feet between us, viciously backhanding me with a blow that sent my head flying, I'm sure I might almost have convinced the Sheriff to let me go.

But fate would not have it.

The explosive intensity of my father's familial rage couldn't have been faked.

After I hit the wall, Sheriff Ball nearly had to jump between us, so that my dad didn't throttle me right there on the spot.

I got up. I was kind of woozy, because my head had actually hit the wall, and my nose had gotten bloodied.

I sat without a whimper escaping my lips, letting the blood drip down my face, wearing it like a badge of honor. I wasn't about to wipe it away.

My dad lunged for me again, as I continued to ignore him.

Sheriff Ball looked sadly at me, as if to say "Man, you've got a lot worse problems than me to deal with."

Unfortunately, I agreed with him. My dad signed the paperwork to release me, and without further "urging", I went with him, head down.

A few minutes later, Cherry's mom came in reeking of cigarettes and booze. The alcohol smelled like it was coming out of her skin. I guess with my dad around she hadn't been able to drink for the entire trip. As soon as they had gotten into town, she made some lame excuse to go off to get a bottle. She was damned near crazy from not getting her alcohol fix for three days.

My dad was already crazy. He threw me into the back of the car. Cherry slunk into the back seat on the passenger side. Cherry's mom jumped into the front seat and immediately started chattering.

My dad set his jaw and began to drive, looking straight ahead.

He didn't blink. He didn't put on the radio.

I looked over at Cherry, and she looked back at me, rolling her eyes.

We were both thinking the same thing.

"Damn! Wish we were back in jail."

Off we went, a huddled, angry bunch.

The return trip was an uneventful, although tension-riddled, three days.

My dad looked haggard. He hadn't stopped to rest. We had to sleep in the car on the way, because he wasn't taking breaks, or letting Cherry's mom do any of the driving.

You could say he was pretty wound up by the time we rolled into town. We dropped them off at their house, and he sped the last few blocks home.

He could barely wait until he got the garage door closed, before he dragged me into the kitchen and started thrashing me within an inch of my life.

He hit me so hard with the first blow that I didn't get up for a while. I couldn't see that good, because my left eye was already starting to swell up.

I felt alone, even though I knew my mom was somewhere in the room.

Believe it or not, she was about eight months pregnant with my youngest brother. The two of them could only watch his mounting fury.

My father was sputtering. He wasn't even speaking words at this point. You could hear an occasional grunt as he took a breath in between swings at my head, my torso and face.

I fell into the corner of the kitchen in a crumpled heap, unable to get up, as he was about to kick in what was left of my head.

By this time, two teeth were missing, and both my eyes were blackened shut.

I couldn't get up. Something was wrong with my leg. I couldn't figure out what it was. I figured the next blow or kick would finish me off.

As I lay there, I could feel my mom's huge pregnant presence nearby. She had placed her body between my dad and I.

"If you touch her again, I'll kill you" she quietly said.

Even in my miasma of pain, I was stunned. I've never, ever heard my mom utter such words before, or since. She was dead serious.

I knew we didn't have a weapon in the house, so what she planned to do to him, I didn't know. I think it was the mere threat of violence from a person who'd never been violent before, that had the needed effect.

He was shocked enough to stop hitting me. He looked wildly around, the whites of his eyes showing. His color started to come down from beet red to fiery red, eventually returning to skin tones.

He started to breathe raggedly, still not speaking.

He turned away, unwilling to accept what he had just heard.

I almost didn't survive the beating, but they decided against taking me to the hospital. No uneasy explanations to give, no worries.

True to her words, once my father stopped the beating, my mom said nothing more about "the incident". The family had a tacit agreement, no one would ever mention it again, as though it hadn't happened.

My mom slowly nursed me back to health. Lucky for them, there were no broken bones, but they still had to keep me out of school for a long time because of the bruises.

The school accepted the bullshit story they told them about

why I was absent for three weeks.

As for Cherry, I can only imagine what her stepfather and his wicked cane had done to her. I'm quite sure that once her mom got home and started hittin' the bottle, she was passed-out-oblivious to the beating Cherry got.

I never did see Cherry again.

I don't like to dwell on it.

****

And that's the conclusion of Cami's "Mardi Gras- Done the Hard Way", brought to you by your favorite ration of passion, The Sextel Network.

* * *

The beginning of the next day-cycle Frank went wearily to his last psyche out-take evaluation, exhausted and somewhat unfulfilled after the actual deed had been done. Even though he had surely killed The Milliner, his long sought after goal now seemed an empty mockery.

Is this all there is?

He shuddered to think of his future, now that all ties to his past were gone.

3

Anti-Vigilante is Born

They'd Rather Be Right

oh how i ache for you,
you,
isolated in a dull gray cloak of logic,
respectively seeking (no) meaning in life.

for,
if everything is determined already
and we have no say to delusions of accomplishment,
nor any way to deter strife,

we are but puppets
strings hanging limp,
no movements to make of our own.

then, there it is.
everything is decided for you-
taking your reason to be,
leaving to your empty imaginings
a mind that will never be free.

you may have
the answer to everything,
but you leave the wonder,
the mystery,
the beauty of life
mutely knocking at your door.

- BY THE BARD, PHOENIX GAET
(IN RESIDENCE AT WHITEHALL)

Cami is OUT, AV is IN

Auntie Vee, being one of the few select patrons (numbering in the mere hundreds) of this famous Flying Dutchman hostel, was becoming irritated at the servobots hovering around her solicitously. They fairly hummed in their antsy anticipation of her every culinary whim. She had asked if they would silence that horrendous noise that passed for music, but to no avail. It was, after all, a required WC run broadcast.

She prudently waved them off with one perfectly manicured hand. Picking up the authentic plastic menu, she pretended to read it although she knew exactly what she wanted.

Wishing to frustrate the bots for a little while longer, she languidly gazed at the other patrons, taking in every detail with relish.

She was intending to download these new mems into her main interface later, and they needed to be crystal clear.

The two techies in the booth next to hers were earnestly discussing the business of OTEC, their engineering jargon at once familiar and comforting.

One man had a shock of white-gray hair, thick and unruly, with a single part on one side, which was determined to go awry, no matter how many times his hand brushed it to the other side. His bushy eyebrows waggled as he emphasized his main convo-points.

"But, you know it's common knowledge that increased (1) 'cavitation performance is a prerequisite for economic tidal turbine applications.' (pg 17) This is a textbook case," he insisted.

Vee seemed to remember the man's face from her briefings with the personnel department at OTEC, and tried to bring up his name. She knew he worked at the Rankine Generator plant.... hold on, she had it. Baru Fundy was his name. She sat back with a little mental grunt of satisfaction and tuned into their conversation.

Fundy's light blue eyes glinted with humor as he realized that he had inadvertently started "A Discussion" with his colleague. He kept the smile from creeping into the rest of his features, and looked down at his blue skin suit overalls, dusting off imaginary crumbs. He was daintily munching a toasted cheese sandwich (an exotic Olde Earth dish), toying with his ankh necklace fashioned out of faux-ivory, which completed his ensemble.

His friend was sporting the latest fad in bodmod (body modification), appearing to be very visibly in the eighth month of pregnancy. It was hysterical; the man actually waddled when he had walked into the hostel!

"Baru how can you be so pigheaded, you know there are (2)'various trade-offs between efficiency and the cost of turbine regulation.' I myself conducted the research on 'variable pitch, versus fixed blading, single versus double regulation and the use of variable speed turbines,' (pg 43)" Severn returned earnestly. He continued what appeared to be an old ongoing argument between the two companions.

Vee observed that they both seemed to be enjoying the debate. She began to feel more relaxed as the rhythm of the debate began to take shape. Baru was the aggressor, keeping Severn on the offensive, who was still trying to establish his first basis for argument.

"Why, (2) 'coupled with rectification and inversion to achieve maximum output, the power system frequency' (pg 43) could reduce the cost of all the rotating machinery!"

Severn turned his bald, milk chocolate brown skinned head in search of a servobot to fetch another round of drinks. His was lemonade and his buddy was drinking a particularly toxic concoction from Olde Earth, called coffee.

He hoped Baru wouldn't notice that his new little Hitler mustache was tickling him abominably. He'd be dished before he let on, wanting to avoid a little lecture from Baru. He smiled at Auntie Vee with a polite once over and dismissed her with his eyes. He turned back to pursuing his next salient point, now ignoring her completely.

Peering over his gold rimmed John Lennon glasses, he noticed that his friend had stopped listening to him and was applying himself to his sandwich. He in turn attacked his sushi and Pad

Thai cereal, pondering his next counterpoint.

As their conversation waned she could hear the Panatechies two tables over, discussing their work-a-day world. It calmed her to hear them talking about their little lives and concerns, a far cry from the problems she and Wad Medani now faced.

TECHIES ARE BAY BAYS TOO

Rance Darrieus and Salford Kapeller worked in the polycelluloid-filter department at the Melissae Hive's detoxification plant, one of the larger divisions of OTEC. They were at mid-cycle meal, in the middle of another day-cycle shift that was looking like a tedious copy of yesterday and the day before that and the day before that.

Rance had been assigned the cleanup of a particularly nasty biofouling of the lower intake pipes leading into the desalinization plant. The problem had occurred when one of the ROV's operating in the Southern Ocean Dome went haywire and rammed the Dome, causing a leakage point.

The resulting damage had caused a 20 percent reduction of output in the auxiliary power array, costing their department a comparable reduction in their paychecks.

"Jack-it, I can't afford the loss of chedits, I'm already overdrawn as it is!" Rance scowled.

But before he could go on to belabor the issue, his buddy Salford made a shushing motion with his hand, causing Rance to stumble to a halt.

"The AudioNet is running another story on Anti-Vigilante, shhhhh" Salford sat still as if that would help him get better reception. "They're saying that Anti-Vigilante is a spy for the AGGIEs, but another talking head is saying No, Anti-Vigilante was one of the mucky-mucks from the MOB who went renegade and started killing off some of their less important employees as a warning of a bigger assassination to come."

Salford quirked his eyebrows at Rance in silent question.

"What do you think? Who is this PHD they're calling Anti-Vigilante? I personally think it's one of the UpTowners who got devolved to the lowest Techie level and then got pissed off and said dish it and went crazy and got put into Retainage. That's where he met the Milliner..... and ... and" Salford looked over expecting Rance to finish his sentence, like he always did.

Rance smiled sweetly and let Salford paint himself into a corner, enjoying every minute of his discomfort. But, an even better reason to keep quiet was that Rance had absolutely no idea who this Anti-Vigilante personage was.

The story was out all over the Net, but it was painfully apparent that most of the factual information and details were missing. What was known.... one of the most violent criminals in the last hundred annuations had just been quietly assassinated while under the care of the most heavily guarded corporations in existence.

This was NEWS! Not in the manufactured sense, but the original meaning of the word. This had NEVER been done before.

Unfortunately for Rance, at that precise moment the debate counterpoint from Malmo Dumjon (Guardian#2, Rehabco, Terrax, and an expert in UVP) began. Malmo's newly published position paper "Violence is a Necessary Characteristic in the PHD Genome" had caught the attention of the viewership, in light of the most recent events within Rehabco, Terrax.

* * *

Talking Head#1:

Q. Why do you think PHDs still need to be violent when we have an abundance of VR Violence Programming available?

A. "Although there's no such thing as government, so to speak, the World's Council does provides all our basic necessities: to wit: shelter, nourishment, and protection from the environment," Malmo intoned using the fake, scholarly accent he affected when on national Vidnet.

"The tacit agreement between Terrax' viewership, the PHDs, and the World's Council, is that in return for being cared for, the

PHDs will follow "the rules", as set forth by the WC.

Most of the rules have to do with simple but necessary restrictions on economic lawbreaking, against physical violence of any kind, and of course, the absolutely dire restrictions on the pollution of our intra-hive life support systems.

These simple, straight forward rules are here to protect and preserve the quality of life for all the viewership, therefore the rules are tolerated.

The rules will continue to be tolerated, only as long as punishment for infringement of the rules is adequate to the seriousness of the crime committed. It is crucially necessary that the punishments are meted out impartially.

Most offenders that end up in Rehabco are here for petty crimes, like overextending their chedit limit, falsifying income levels or not earning enough chedits to pay for their luxury items.

After their waiting period, these offenders are assigned the less onerous work assignments, like editing the VidNet, providing comical fodder for crude e-shows, or working in the underground bazaar. These jobs have a high survival rate, although the pay is krunk.

But when a viewer disobeys the restrictions against physical violence, they are immediately sent to Rehabco, where they stay one Wheel of the Year, awaiting their new work assignment. These work assignments are the dirtiest, meanest, most physically dangerous jobs on the planet.

For the most serious offenders, we assign the most horrendous work; cleaning up the nuke sites; piloting shuttle's of UVPs to the penal colony on Phaeron; cleaning the equipment that vents toxins Planetside, or scouring the human remains (from the inhabited SadSacks) off the Harvesters.

No one in their right mind wants to be a assigned such work, but it's common knowledge that certain BayBays will continue to perform acts of senseless violence.

Why?

Even the brightest of Wad Medani's MM's cannot figure out how to genetically engineer all the violence out of the PHD race."

Talking Head#1:

Q. But, is it in fact a **necessary** component of our very beings?

* * *

Auntie Vee was listening to the broadcast with only half of her attention. She had heard Malmo's spiel before and had found it rather dull the first time. Her mind wandered off for a moment, and she decided to use this down time to "record" this morning's performance by her beloved dance troupe, The OSMOs.

* * *

Closing her mind to the cacophony around her, Auntie Vee "tuned-in" to her favorite aerial acrobatics team, now performing in The Aviary. They were beginning the "Whirling Dervishii", an intricate aerial dance in triplets. Like Space-5 chess, it involved precise Dissemblage and Assemblage point coordination.

The performers dressed as robins and wrens, swooping gracefully in and out of formation. It was very relaxing, almost hypnotic, to watch.

If you half-close your eyes you can imagine they are birds, dancing an ancient ritual of introduction. Today the troupe decided to make the performance more interesting by releasing a benign aqua-colored gas into The Aviary. Dancing in the gas would lend the dance an ethereal quality, but it also made the performance more dangerous, because it was harder to see the other dancers.

The leaps and flips would have to be made with pin point accuracy; else the performance would turn into a bloody disaster.

Of course, this was part of the appeal for some of the viewership. Especially the Hutu men, a small but vicious enclave of the MOB. They reveled in the idea that they may see someone die today.

But for the most part, the rest of the viewership authentically appreciated their art form and wouldn't want any of the dancers hurt for any reason.

As a dabbler in genetic engineering, Vee had been quite successful in helping some of the dancers gen-ennge vestigial wings onto their shoulders, aiding them with lift for the power moves.

You could say, she had a stake in their remaining uninjured, because she had a large holding in the OSMO'S (or self-proclaimed Oscillating Osmosiers). She had conned some of her MM friends into engineering a skinsuit that incorporated opto-electronic oscillators into the fabric of the suit, allowing its wearer to echolocate any moving object.

This way, the dancers had a modicum of extra protection against accidental collisions, because their personal GPS systems could locate any other dancer in the dome.

Of course, since the dancers ran the risk of gross miscalculations of another dancer's moves, accidents could still happen. But these dancers were at the height of their physical athleticism and had trained for day-cycles on end for this performance.

They were definitely ready, in their most excellent physical form and state of mind.

Vee beamed with pride at her little flock, knowing that they were giving the performance of a lifetime.

* * *

Snapping back to the other monitors, Vee realized she hadn't missed much, as Malmo ended his lecture. Her food order had arrived, and she anticipated every sip of her albondigas soup.

She kept her mind on the present nano, knowing that on the way back to her private labs, she'd have time for a private chat with Wad Medani who was stationed on the other side of the world.

Usually she wouldn't have dared make such a choice, to call him from the semi-public comm lines in her 'mop, but she knew that the VBB was getting ready to broadcast one of "Randi's Randy Bedtime Stories", and there would be absolutely no one interrupting their call.

* * *

Welcum, to another sleazy session of "Randi's Randy Bedtime Stories."

The Prepaid Acid Test Complete With Feathers And Head Band

His ears changed quickly, if you caught the pre-change twinkle in his skin, you knew (or could guess) what was coming from the corner of mind's eye. Perhaps I should have been more cautious in dealing with this pixyish sprite, yet, being cautious would have lost me an adventure with the Prince in His Cave or as I fondly remember, the Prepaid Acid Test Complete with Feathers and Head Band.

I've actually forgotten how we met, and it really doesn't matter, for in this particular case the middle of the relationship (it lasted one whole day), was far superior to its beginning or end.

He was a student at the college I was attending, and we met near campus.

"Hi, haven't I met you in such and such a class, party, orgy, etc.???" he asked.

Our formal introductions now over, I casually threw him a tester line.

"You know, I'd do just about anything for a hit of acid right now."

I grinned a maniacal grin at him, which he found quite charming. He grinned back, and we had an instant understanding that for an exploratory evening in his personal stash, I would in turn love, fondle and ball him for the rest of our evening together.

Sexual adventures need supplies, if you do it up right, so we went shopping next. Stopping at the headshop on the way to "there", he bought three small bottles of sweet body oils. He fancied himself quite the dabbler, you know. He promised me ticklish fun with a waggle of his expressive eyebrows.

On to the grocery store, where we picked up a bottle of wine, a chunk of jack cheese, half a dozen eggs and tortillas, then hauled it all back to his banged-up truck. While walking through the parking lot, a thin, well dressed black man standing next to a big Lincoln asked if we'd like to buy a million watches (he wore about half a million on each arm).

We declined as a couple. This action affirmed that to the general public we were to be viewed as "being together". It was then "understood" that I was to be affectionate with his ears, butt and hips respectively--in and out--of crowds.

He was cracked-up in his own little world, thinking himself untouched by the opinions of others. Most of the time I didn't talk, letting events flow along, watching his mood and body. We both knew the part I played.

YEEEEECH----- you might be saying to yourselves. How dare you!

Here was a princely pixie, an innocently devilish looking man with soft hideaway horns and sharply pointed ears, the unfathomable depths of blue-green merriment twinkling in his eyes.

He introduced himself as the Prince of the Cave, and I was very willing to open myself to his bidding, his pleasures and his fantasies.

When we arrived at his rundown shanty, he remarked lightly of what the neighbors would think. I scoffed at their gossip and carried in the groceries. Tinkering around in the kitchen I scrambled eggs and folded us up some burritos.

We went through the ritual of feeding each other, which is sloppy and funny. We smoked some ganja pot and drank fruit flavored wine.

He had to finish his homework for the next day, so I made myself cozy, checking out a book of his on various positions for love making. I was giggling as I rolled lots of his good grass, and began to get naked, slowly, as per instructions of the Prince.

While the acid worked its way into my brain a little at a time, I saw that the walls of his cave were covered with posters of East Indian artwork. We decided together to dedicate our acts of ecstasy to the various Goddesses and Gods whose pictures adorned the walls of his bedroom.

The Prince, homework finished, pranced over to the foot of the bed, headband holding his tiny goat-horns in place. With that strange grin on his face, he sat down and started dabbling me with sweet body oils; coconut and warm cinnamon spice.

He giggled as he came on to the acid.

Taking him by the arm, I led him to the bed and began lavishly massaging him with the oils, as we imbibed in more drugs. After a leisurely time spent suckling and teasing his cock, balls, tummy, nipples and mouth, he began to slowly come onto and inside of me.

He could get it up more times than anyone I'd ever met! We romped all over the bed playing master and frightened slave girl (this was one of our themes for the evening.)

We moved onto naughty maid and stern master, then pirate and randy wench. All things considered, I pleased him.

"OH........MOMMY!", he cried out (with the strangest expression of orgasmic joy I'd ever heard), then lay there limp on the bed.

I felt like a voyeur in my own bed, and a little bit guilty, not being altogether mentally there with him. You see, as the acid was surging through me, my mind was alive with scenes of Jeff, my current boyfriend, making love to me.

I could feel Jeff's hands on my face, tracing his way down my body, caressing my arms and thighs. I could imagine his delicious, meaty cock pushing into me, growing thicker and longer with every thrust of his hips.

When the Prince finally "came", I was very much an observer. I felt distant from him, and at odds with myself.

After he fell asleep, I waited patiently for dawn to bring back the toad that he would surely become.

But in the morning, I was delighted to recognize him as a

man, although somewhat drawn and pale looking. Not much was said on the drive back to my apartment.

The acid had been good. The sex had been good. We'd made a fair trade, I think.

As he stopped the truck to let me out, he silently handed me a beautiful choker he'd made for me sometime while I slept that evening. The choker had deep purple feathers and wooden beads worked into the band of black velvet, clasped in silver. Unique and stunning, it was a rare gift.

He drove off without another word. I never heard from him, nor saw him again.

I had met a prince in his cave, strewn with old books and well worn rugs.

Warm colors, sudden greetings from strangers, and fleetingly illusive perceptions still bring back memories of our unique encounter.

* * *

sobriety sucks

I woke up this morning, early and bright
I've been sober 24 hours, I'm trying to get it right
I've taken all the steps and chanted all the prayer's
I've witnessed and I've spoken,
but reality's a bear.
I tell you once, yeah yeah
I tell you twice,
it may just be an herb to you,
but it's my vice!
sobriety Sucks
yeah yeah yeah
sobriety Sucks

By Kayla Raconteur

Malmo Dumjon here, reporting to work as usual. Yeah, I'm

sittin here nice and comfy, jammin to the latest mix by the Masta Mixa Pavarti. I've got my kinnick- kinnick rolled-up into exactly 10 jays, one for every watch during the midday-cycle, since I gotta be here anyways.

Feckit, if the smoke alarm in this damn tin can goes off, I know how to disarm it so's they'll never know it went off. I am fully tut.

I love that I can sit here watchin my mega-multi-media vidscreen for free, and, that I have three of them. Yeahhhhh, one is a reg'lar vidscreen with WC sponsored 'gramming that everyone is "encouraged" to watch. The other two vidscreens are for work; one to monitor my Retainees, and ones filled with the usual MOB sponsored drivel.

I'm lazy, not a dumb squeeze like Yosaka who's in charge of monitoring his Retainees, as well as off-planet surveillance of all the Retainees of Rehabco, Phaeron. Sucka.

My preferred work assignment slot is the midday-cycle because by then every one of the Retainees on my floor is asleep. That'll give me time to spare to rehearse some of my killing moves for the upcoming debate tonight.

Yosaka, my opponent for tonight's debate, prefers the midnight to morning shift. He's got a worker-bot mentality, and having just come off his shift, he'll be none too fresh for the debate.

Ha! I already have a slight advantage over the senseless git.

The Office of Genetic Engineers will sponsor the debate, today's topic being;

"Is Violence in the Human Genotype Necessary?"

I am taking pro and Yosaka is taking con. Naturally.

Being Historians, we're both vastly qualified to speak on the subject, having studied ancient Earth history for many annuations. We Guardians love history. And because current affairs are being covered by the MOB and TDM nanosecond by nanosecond, there's no real sense of gaining information or knowledge by studying current affairs.

In order to be able to join the ranks of Historians (the elite category of PHDs who are allowed access to the most ancient entries within the TDM data files); you have to get authorization from Grandpa and the Calava Corp. gen-ennges who developed the original machine.

You must prove your seriousness of mind and purity of intent in the acquisition of this knowledge. You must swear that you will never reveal, copy, reproduce or distribute any of the 'gramming you access.

After synthesizing and experiencing the feelings and thought patterns of hundreds of these ancient beings, I came to realize that violence is inherent in the human genome. It can not and should not be deleted out of our programming.

I observed that it is inherent in both genders – there were only two at that time – male and female. Not only did they engage in violence against their own kind, but they extended their violent predisposition upon the very planet they lived on, against the earth, plants and animals living there.

There were different expressions of that violence; when countries engaged in it they called it War and made up rules on how to "win" it fairly; when it was within the borders of a country (a form of organized societal violence), they called it gang war and the local officials fought against it.

Lastly, they engaged in "interpersonal violence", which ranged the gamut from people beating up, raping and torturing each other, to outright killing each other.

For some strange reason, the Ultra Violent People were punished by society for their interpersonal killing, as opposed to the publicly rewarded, governmentally sponsored violence (War).

The common people were encouraged by their governments to support the official, government sponsored Wars (they called this patriotism), and if the common people opposed the War, they were considered unpatriotic and duly punished.

You can look at the utility of violent behavior in two different ways.

On one hand, humanoids were at the top of the food chain

and they were multiplying voraciously. There weren't any natural predators to control their numbers, as there are for every other animal within the natural order of things, or within an ecosystem.

In all of nature, there is a match of predator to prey; everything has a stop-gap, in order to balance the population of any one particular species. This maintained a perfect balance within the ancient ecosystem that they generally referred to as Nature.

Species would then interact, fighting each other for dominance over territory, sexual mates, and food sources. Society's human beings evolved into what they termed "civilized society", in that they developed structures to live in, weapons and tools to improve their living conditions and eventually, technology to enhance communication and the spread of their own culture.

In other words, the humanoids weren't living as all the other species of animals did that lived within Nature, but instead lived in constant opposition to Nature. Humans desired to enhance and control Nature through technology.

And, even as there were fewer and fewer individual countries and individual lands to fight over – even as humans conquered the entire natural world or attempted to rule and control it – they still continued to fight with each other over the few natural resources that remained.

The very last of these epic fights (Wars) involved the entire human race. By that time, there wasn't any more open land to fight over, the natural resources had been drained to the last drop and the food supply was running desperately short worldwide.

The only thing worth fighting over was the right to clean water, because at the rate of population growth, there wasn't enough left for people to drink, much less to water their crops.

The World Water Wars (referred to as WWWs) was the last War on Earth, and it nearly destroyed them all.

Even after the carnage was over and the bombs' smoke cleared, people still fought each other on an individual basis. They especially hated each other for differences in religious beliefs; they were willing to fight to the death to prove that they were right and the other group was wrong.

The World's Council was formed out of the ashes of the WWWs, and the new society's belief system was based on the premise that "emotion is the root of all evil."

Due to the fear of inherent violent tendencies in the population, the Genetic Engineering Office, formed in the early OOs, had come to the decision that they must remove the genetic disposition of violent behavior in humans.

They announced that they made this decision "having only the good of the people in mind." After all, theirs was a respectable, established scientific profession.

Genetically modified food had been growing for decades by that time, and had not proven harmful in any way. In fact, it had provided abundance where there had been famine.

Within a 20 – 30 year time frame, genetically modified organisms (GMOs) had not proven to be the "Franken Food" that the original scare promised. The Gen-ennges had miraculously created strains of food resistant to drought, resistant to saltier soils, resistant to the harsher, more toxic climates worldwide.

Once involved in agricultural products, the gen-ennges knew no boundaries. Soon, the idea of using human genetic engineering to improve the species came under their microscope.

As far back as Olde Earth 1997, Scotland's Roslin Institute announced that they had cloned a sheep named Dolly, using a single cell of DNA from an adult sheep. (*)

The cloning was in its infant stages, being accomplished by extracting genetic material from the cell of the animal that was being cloned, and placing it in an egg whose nucleus has been removed. The egg was then mechanically "activated" to develop into an embryo. The resulting embryos were transferred into a surrogate mother.

Soon after Dolly, came "Snuppy, the world's first cloned dog", who was created in the Seoul National Laboratories. (**)

Then cloning became all the rage, and a host of other cloned animals followed including mice, a cat and a horse.

The next phase in genetic modification came with therapeutic cloning, which involves creating early embryos to use as a source of stem cells. These "master cells" are capable of developing into any cell type in the body. Stem cells are harvested from the early embryos, which were used in research that resulted in the cure of diseases like Parkinson's or diabetes. The stem cells were also to be used to treat spinal cord injuries, cancer and heart disease.

It seems that Dolly the Sheep's makers and their "copy" cats, had set the stage for the next step toward the complete acceptance of human genetic engineering. The gen-ennges wisely began their experimentation first on plants, then animals. This slowly, insidiously led to the use of genetic engineering on humans.

The Genetic Engineering Office (GEO) found they were able to maintain the quality of the test subject's lives, allowing them to enjoy a healthy, natural span of life.

Inevitably, when the WC announced to the viewership they had cleared the GEO to genetically engineer human organisms, it seemed as commonplace as DNA testing.

Soon, the viewership got personally involved and "gen-mod for your baby" became the latest rage. A couple could now program their nestling, choosing; gender, physical characteristics, or specific traits, like the propensity for musical or mathematical genius.

That option backfired in a big way. As generations of people began copying each other's 'gramming, there were masses of people with similar personalities. These masses began to form "waves" of PHDs so similar in their traits, that the WC and the newly forming MOB could actually forecast the viewership's exact behavior patterns.

Specifically, their spending patterns. It was known as "mirroring the customer".

That's really when the Media Overlord Barons started to become a power to be reckoned with. They began as media moguls, hungry to scrape whatever money they could out of the bored viewership being held hostage by their limited environmental surroundings. The viewership could not leave the Warrens, as the air above ground was too toxic to venture out into. Besides their editing duties, they had nothing but free time to fill with entertainment. Which suited the MOB.

The media barons tracked groups of PHDs who were genetically engineered with a certain set of assigned personality traits, and profiled their purchasing habits. Before long, Asna Janghaghir, the main MOBster, put forth a bold action plan to implement their newfound knowledge.

Rather than trying to create original programming, and attracting viewers to them, rather than attempting to sway the slavering masses, he suggested that instead the MOB GenMo their legions of viewers to desire certain 'gramming.

"Hey, it's like we're pushing the cart from behind, instead of waving a carrot in front of the horse who's pulling the cart," Asna Janghaghir proclaimed, from his absolutely safe position as the fattest cat lazing on top of that cart.

In one bold marketing stroke he suggested that they directly influence their audience, in essence "causing" their viewers to purchase MOB sponsored 'gramming.

The MOB decided, why not? Would you rather try and gamble, throwing out random ideas, like tossing a huge net out in order to catch a few fish? Or would you rather have your fish pond stocked?

Unilaterally, the Board Of Trusted agreed that the MOB would legally sponsor GenMo 'gramming for a select number of Nawals. There were quite a few elite PHDs who wished to modify their children, but needed some financial assistance.

In return for the MOB sponsored GenMo, the meshlings agreed to allow certain genetic traits to be instilled in their nestlings, in order to reproduce the perfect MOB viewership qualities.

PHDs choose GenMo for several reasons: a gender preference, a preference for specific racial characteristics, and a propensity for unique characteristics, as I said – musical or mathematical or creative genius.

In all this genetic stumbling around, what with removing and adding characteristics, they also gen-mo'd out the specific genetic weaknesses causing disease. Disease has been eradicated. Instead of trying to "cure" disease from the outside, either by ineffectually attacking the symptoms with drugs or by trying to keep you as comfortable as possible while you are dying, the gen-ennges simply deleted the genetic programming that creates defective cells.

Just as they genetically modified diseases out of food crops and microorganisms, for that matter.

Genetic Engineers had been developing entire viewership populations by the time the WWWs came around.

With regard to violence, even as late as the 30s, the World's Council scientists working on the Human Genome Project could not modify, strain out or otherwise remove violence from the human genome.

I know. Yosaka has had access to the same info and 'gramming, but he came up with a different conclusion than I did.

Yosaka is going to debate tonight on why people are violent in the first place, what I call the "whys" of it all.

"Oh, the hell with that!"

Why people are violent doesn't mean shit to a tree. It's like a dog that's barking in the middle of the night.

Why is the dog barking? Does it really matter why? The solution is, shut the damn dog up. Right?

In my opinion, it really doesn't matter why they couldn't strain it out. For some reason, violence is an unmodifiable component of the human genome. Period.

The interesting fact remains, human beings inherently have two traits you can't subvert, two that can't be Gen Mo'd out of the gene pool; those being curiosity and violence.

The bell intoned designating the end of his time to lay out his first premise.

Malmo's mind raced on, thinking quietly to himself, as he ignored Yosaka's fumblingly clumsy beginning of a rebuttal argument to his first premise.

Of course, I knew that once my MOB bosses acquired dominant shares of the Calava Corp's. GenMo Labs, they'd approach violence in a different way.

We figure that allowing random violence is bad for business. The UVPs were killing off our carefully cultivated viewership, and we seriously depend on them to buy our 'gramming.

No. No.

Random violence isn't at all valuable, as far as the MOB is concerned.

I believe that Asna Janghaghir asked Bepko Da Silva, his Primo Gen-ennge, to address the problem, and this is what he did.

Da Silva came up with the notion that, rather than us try 'n siphon out the violence, why not concentrate it in a few chosen PHDs? This way we'll own the genetic patent on the nestlings, (who will, in essence, "contain" the violence in the human genome.)

After intensive UVP training and 'gramming, we'll release them into the general population to do their worst. And, we'll catch it all "live" for all viewers to jack-in to, scoring the highest RQs since the Senselles arrived.

By Vee, we'll best the Senselles at their own game!

Of course, this is a secret of the "if I tell you this, I have to kill you" variety. A huge, monstrously ugly, well-kept corporate secret. One the viewership is never to know.

The Calava Corp.'s gen-ennges know, and are being paid handsomely to keep their collective mouths shut.

It's true. They've known all along that the MOB is responsible for releasing these creatures into the general public. They know the MOB allows the UVPs to visit unbelievably savage violence upon the blessed viewership, without any rhyme or reason.

Suckas! The viewers never see it coming, and once released, no one is able to stop them from completing their 'gram. Even my MOB masta, Asna Janghaghir, cannot stop one, once activated.

It is whispered that Da Silva, whose name is second in evil only to The Great Fear Itself, began as a gen-ennge for the MOB, then joined Wad Medani's MagicoMathematicians, after the krunked-OUT Senselles arrived.

Curiously, he's one of the only PHDs who've been allowed to join the hallowed ranks of the MM's, because Wad Medani flat out doesn't trust us. (As well he shouldn't.)

Wad Medani knows that Da Silva is brilliant. Dangerous, but brilliant. And, that he can think creatively around our severe limitations. In this blasted place where we're so limited in raw materials, Medani knows he needs a guy like that around.

Why do I say "us?" Because, I work for the MOB. The MOB, being the procreators of the UVPs, is responsible for them, and so created the system of Retainage, where PHDs are housed when they break the rules of society. Retainees waiting for their sentencing in Rehabco, Terrax live comfortably, without inhumane or derogatory treatment of any sort, until they have been given their next work assignment authorization

A work assignment authorization is similar to the Olde Earth barcode system, it's basically a holographic date-code strip that's superimposed into the inside of the left wrist.

Since it isn't physically there, it can't be physically removed. Genetically programmed to disappear after your work assignment time period is over, each data strip is encoded with an assignment or work level permissible, which is according to the level of impropriety committed.

Once you leave Retainage, you cannot make a living, except at the assigned work level, until your assignment is up.

For example, let's observe the lesser offense of economic disparity, such as when a viewer is in debt beyond their ability to recompense the WC Chedit System.

Imagine, your average BayBay living in a mid-level Hive. They have run up excessively high charges for a bunch of 'gramming requests, or perhaps designer drug-shots, but they don't earn enough chedits to pay for it all.

The System knows when One Is Not Paying Their Way.

First time offenders are given a pity party, as far as I'm concerned. All the services and utilities to their cubby are turned off. They are devolved, removed from their cush Hive and reassigned to living in the Warrens, DownTown.

The majority of BayBay PHDs live like this, always clinging to the borderline of debt, but in our equal opportunity economic society, all viewers have the capacity to earn their way back UpTown.

Once devolved down into the Warrens, these BayBays receive a shabby WC cubby barely big enough to turn around in, supplied with second rate prot-shots and thin skinsuits. And that's where they'll live until they earn their way back UpTown.

Usually, it's enough to keep them from becoming repeat offenders.

The Warrens were originally housing units built for the Olde Earth humans to live in after the WWWs. The remnants of human society who had survived that war realized they'd best live underground, because the outside environment had become a deadly toxin. Animals, plants and human beings were all dying from the toxicity in the air, soil and water.

But I digress.

The real reason the MOB created Rehabco was to house all these Ultra-Violent PHDs that we'd genetically engineered, until we figured out what to do with them.

The MOB decided to allow them to wreak a certain amount of havoc, then the UVPs are Retained to await their next work authorization. The UVPs know that they probably won't survive the new work assignments, some are certain to be death sentences.

If you're assigned to work under the Ocean Dome, you'll be working in the OTEC UnderWater Division, under the direct supervision of Arpad Reitan. The Domes not only collect solar energy, but the oceans also generate energy using Wave Energy Conversion machines and underwater tidal turbines to create even more electrical power.

Within the Dome, the oceans are now contaminated with industrial grade pollution. The water cannot be cleaned for human use; it's turned into a big soupy battery, filled with heavy metal sediment, corrosive chemicals and fluids. The only machines capable of surviving this environment, are the ROVs, the Remote

Operating Vehicles, and they can only last about one-half of an annuation.

The ROVs have to be continually maintained, because you don't want the bastids to break down. Since they're remotely operated, when they breakdown, they have to be manually rescued from that hell of an ocean.

This is where the assigned UVPs step in, diving the depths of that abyss, braving the toxicity, facing the scatterstorms that are raging inside the dome. It's one of loneliest jobs on the planet, working solely with machines, and they are meted out to only the most violent of the PHDs.

Some might question why PHDs would acquiesce to do this deadly work. Why don't they just refuse to accept their work assignment level?

Remember, this is all that the Big Four will allow them to do to earn chedits, and without the work, they'll get nothing to eat or drink, receive no clean air, water or entertainment. Even violent offenders need the basic necessities to live.

Believe me, I've seen it work. The instinct to survive, to live, is too strong. Most PHDs won't starve themselves. Survival's one of the things that we're all naturally genetically engineered to do.

Oh, they may make a feeble attempt at living without chedits to buy their prot-shots, but then their anger kicks in and they realize they don't want to starve to death.

They realize the WC is serious, and will strand them without utilities or a place to live. Once they are shut out of a cubby, they'll have nowhere to go but OUT – OUT into the elements- AboveGround!

Everyone knows if you go OUT, you'll turn into a puddle of goo within nanos.

Nobody can survive that.

That instinct to live, to survive, is another main reason violence cannot be excoriated from the human genome. It is the ultimate survival instinct. If a PHD is physically threatening your life and you do not respond with physical violence, you will die. PHDs must retain their ability to be violent, in order to live."

Mentally ending his debate with a flourish, Malmo continued his personal train of thought to its completion.

What the viewership doesn't know, and cannot guess, is that the worst of the offenders are the UVPs who can't be controlled, the ones with a strain of abstinence so deeply ingrained, that even the threat of starvation will not make them acquiesce to their work assignments.

These have been few, only about 15 in the whole history of Retainage, and we've shipped them off-planet to Rehabco, Phaeron. Phaeron is the satellite moon of a desolated planetoid known as Al-Romaithi.

The terrifying truth is that the MOB sends the UVPs there because even we don't know how to control our creations – they've become our own private army of monsters.

Tearing his mind away from this vein of thought, Malmo checked his RQs, which were moderately pleasing. He scanned his Retainees for any disturbances in their life stats, and snuck a peek at Frank's cell, watching the hated little weasel, who was in Yosaka's care for the last time .

Malmo was going to enjoy following the little runt until he messed up again, and losers like Frank always messed up again.

Malmo was gonna personally follow Frank to the next AGGIE meeting, and bust him right back to Retention.

His MOB boss had promised Malmo that next time Frank was in Retention, he would draw Malmo's cell block.

Then life won't be so sweet for little Frank/ie boy. Oh, nooo.

Malmo grunted softly in anticipation.

* * *

MEANWHILE ON THE FLIP SIDE OF THE HIVE............

This is Wad Medani, checking in briefly from the other side of the world with my techno report for the mid-annuation cycle, or as the Grovers call it "the winter solstice."

Hell. I'm on the other side of the fuckin planet, again.

I miss Vee... I know she'd never believe me if I ever admitted that.

At any rate, when I got here, there was already a Level 1 Enviro-Leakage damage report coming in from the Atlantean Coast Dome. Turns out, this leakage has been caused as a result of that damn ROV that went off kilter.

"We" have no idea why it went off. Unfortunately, I didn't even know about it until they finally managed to stop the thing from creating more damage. For some strange reason, it kept ramming into the side of the Dome until it created a crack in it. Well, it's not really a crack, per se, but a kind of vibrational seepage that was allowing the deterioration of that particular part of the Dome's structure.

And that's not a good thing.

Of course, the particle field that composes the disteldt shield is supposed to be impenetrable. But there seems to be a distortion of the energy material, as if the phantom energy (P-energy) were traveling between particles, pushing in an outwardly spiraling direction, changing the very shape and nature of known matter.

Particle physics aside, it's bad enough that we had the enviro leakage. Originally, when the enviro leakage was detected in the Ocean Dome, it alerted the toxbots. Because there was environmental clean up alert, Safety was also called in.

All the techies that were working with the ROVs (remote operating vehicles) were called in because the thing was out of control, and once they dismantled the receiver for the bot, then it immediately started to sink to the bottom of the ocean!

That jelly-like bisque of toxic jelly, more battery acid than water, contains mercury, titanium, volatile organic compounds, petrochemicals, ash, soot and nuclear waste.

We contain it in Domes because now it's as dangerous as the outside air, however it is extremely useful in that it acts as a gigantic battery, storing the power generated from the Hive and the Dome's shimmering PV collecting materials.

Once the ROV sank, they needed an expert to retrieve it and sent for Arpad Reitan, an off-duty HORSE, and the best of his ilk. Luckily, he hadn't pulled a space assignment, so he was in port.

It's damn dangerous work, the worst being that your craft can disintegrate around you before you can actually get anything done. Reitan managed to go down before I got here, and retrieved the ROV without further damage to the Dome or the unit.

The toxbots had peeped an alert to Vee's personal comm link and she'd sent her two best Enviro-Panatechs, Naxo and Soen, to detox the area. As to the environmental leakage and structural damage to the Dome, they were busily reprogramming that part of the ocean dome's microorganism to heal itself.

The OTEC techies from the Biofouling Div., Iakapana and Olul, were assessing the environmental damage from the toxins that were seeping into the ocean.

My primary job is to figure out what in the world the ROV was doing transmitting outgoing signals off-planet. The signals can't be decoded, so I'm downloading the data strands (all the information it was sending off planet), into the latest encryption code-breaking software.

Even though I'm the only one with access to the downloads, it is appearing in the secret TDM activity files for covert info streaming. I've already assigned my top astrographer to figure out the coordinates of where the transmissions are going.

This is so wack. Just from looking at the raw data, it's pretty obvious that they are being beamed to Phaeron, Rehabco.

Why in the hell would this little ROV, which is an unmanned vehicle, want to communicate with a penal colony that is completely automated?

Supposedly the ROV programming is tight. It's just a simple retrieval machine for ocean retrieval operations. If OTEC has to do any repair on machinery that's fallen into the ocean, we have to retrieve it with the ROVs. Supposedly nobody but OTEC has the ability to change that programming, and it hasn't been changed for centuries.

So, why in the world would it be transmitting information? This is not an information gathering machine, so that alone is highly suspect. Something is going on that we need to know about.

I know that I'll probably have to talk to Asna Janghaghir personally about it because the MOB owns the Rehabcos –both Terrax and Phaeron. They run Retainage. Their Guardians are the ones who oversee Retainees and assign the work assignments. The Guardians also have unrestricted access to TDM downloads.

This makes sense because they're our historians and have access to so much information that they know best what these criminals should be assigned to do in order to rehabilitate them. Either the returning Retainees make it back into the main stream of society, or they perish. I don't really care which it is.

My main concern here is with the MOB, and whether they're the ones orchestrating this. They claim that the UVPs (Ultra-Violent PHDs) who cannot be rehabilitated are sent on a one way trip to Rehabco, Phaeron.

That is, if they don't die during their horrendous work assignments (ocean retrieval work, cleaning toxic seepage materials, or the retrieval/recycling of raw materials for the Big Four Industries.) The UVPs recycle everything; they work the recycling plants, the reclamation centers, repair the Harvesters.

It's common knowledge that the death rate in Recycling is a guarantee. Most don't last even a semiannuation.

If you persist in being violent, you're sent to Rehabco Phaeron, to live out your short, bitter life with automated robots. As far as I know, Retainees are always on camera, being monitored at all times; from their life signs to what they're thinking, what they're eating, and what they're doing.

At least, that's what the MOB assures us is going on out there. Hmmph.

This is why I need to talk to Asna about what kind of control he really has on Phaeron. What the hell kind of involvement do they have in this mess down here?

And, I'll tell you what else . . . Vee is going to be livid when she realizes how many people were involved in this before I even got here. Asna's people know, because they're the ones who own Arpad's contract, their recovery teams are the only ones contracted for salvage operations both in deep space and deep sea.

Much as I hate to deal with him, I've got to set up a meeting with Asna right away.

* * *

Grist for the Mill

You offer me nothing
come on, don't follow me.
You've got the will to fight,
trust your own heart.
My life is my own,
you'll not take my yesterdays,
I won't depend on you,
time is mine.

By Kayla Raconteur

Cami was kickin back, pondering her current situation and wondering how she had enmeshed herself once more in the political intrigues within the upper echelons of the Senselle organization.

Ambivalent about politics, she nevertheless felt loyalty both to Auntie Vee and Wad Medani, who were at odds on certain political issues, but not on their singular vision of world abundance for the viewership.

Cami was torn between loyalty to the Sisterhood, who'd given her everything that's good in life; creature comfort, security, a family who would never turn their backs on her, and her own "adopted" sister, Damaria, who was often at cross purposes with the Sisterhood.

On one hand, she had deeply believed the Sisterhood was her world, but then there was this nasty business with Frank/ie.

If the highest order of the Senselle organization could agree to use an innocent person this way, what would they do to someone who openly disagreed with their goals?

Wad Medani had been like an uncle to her, he always had time to listen to her problems, even while taking the weight of the world on his shoulders. But in this, he wouldn't go against Auntie Vee, they were in it together, for Vee's sake!

In fact, when Cami asked Wad Medani about seeing Frank/ie after her recuperation was completed, he didn't give her his usual straight-forward explanation. Instead he snapped back an uncharacteristic, "No, just leave it alone!"

Cami thought that since she had been asked to rescue this person, she had, at the very least, a right to see what they had done with her.

Or whether she'd even survived the process.

Ever since Wad Medani had shut her down with absolutely no consideration of her stake in the matter, Cami was itching to figure out a way to get through the obstacles that had been put in front of her.

Whenever she got into this particular bent, she began to feel angry. Then she'd allow herself to get lost in the past, in order to remember how to do something she'd forgotten how to do.

She needed to remember how to use anger to fuel action that would precipitate change.

Digging way back into my past, I remembered a time in my life back on Olde Earth when my anger was put to good use. That's when I began to understand how to become a rebel with a cause.

* * *

You Can Fight City Hall and Win!

After high school, I went through one year of college away from home. By the end of the year, I didn't see how carting around 80 pounds of books was going to help me earn more money, so I tried a trade school.

I chose South Bay College of Baldwin Park, because I could get in with a Fannie Mae government loan. In order to "qualify" for this loan, you must show that you have very little money, and you have to promise to pay it back after you graduate.

They even give you a whole year to get a job before you have to start paying it back in small payments.

I had this one in the bag.

Naturally, being a poverty stricken minority student with no visible means of income, I qualified for the loan.

Hurray, I got into the school!!!

I liked the administrators, and the registration counselors I met on the first day of school. Unfortunately, once I got to the actual classrooms, boy, everything was totally different from the University I had attended my freshman year of college.

The teachers here sure weren't like the ones at the college.

One was a nondescript math teacher, with a completely forgettable name. He'd read the newspaper during our class time.

He'd put his feet up on the desk and tilt his chair back, reading his paper, instead of teaching us.

If anyone was dumb enough to ask him a question, he'd snicker at us from behind his paper and mutter something about "reading the book for clarification". (All the while he was supposed to be answering our questions and teaching us business math!)

Sorry for the interruption, Mister.

It would make me crazy, because the whole point of being here is to get my skills together and get a real job.

Here I am, trying to better myself so that I can have a better life, and this guy isn't even trying to do his job.

Even though he's being paid to teach us, he has nothing but contempt for us. He despises us. You can tell he thinks he's better than we are, just because we're all from lower income backgrounds.

(But, he already knew we're all poor people here, that's how we qualified for the loan program.)

It makes me so angry that since there aren't enough desks and chairs in the classrooms for all the students, some of us have to stand during class.

People end up leaving for the day, right after they've signed in. Who can blame them? Who in the hell wants put themselves through this, when they can't even sit down to do their class work?

When I found out there weren't enough books in my classes for all the enrolled students, I asked my teachers what I should do.

I was told to wait until the new shipment of books comes in. In the meantime everybody else is continuing the lesson, while I'm missing out on assignments, and getting further and further behind.

There is one teacher, Mrs. Johnson, who's a standout mentor, and friend to the students. She's such an angel! She has dignity, and exudes strength and purpose.

Thank God she's the teacher who's teaching the accounting module, because that's my area of concentration.

We don't have computers in our class room, instead we are learning the basics of accounting from scratch. We're learning double entry accounting, all according to the Generally Accepted Accounting Principles.

Mrs. Johnson seems to actually care whether we're learning the subject or not.

She's strict, but kind, and although sympathetic to our problems with the school, she can do nothing to help us without risking her job.

One day, after dealing with this craziness; with the lack of materials; and the teachers not giving a damn about it one way or another, I had had it.

The "school" consists of two small rectangular buildings, each laid out with a reception area in front, then one long hallway in the middle, the classrooms off to each side. The administration building has offices on either side of the hallway.

As I make my way from the classrooms out to the parking lot, I head out to the area where everyone stands around and smokes. I don't smoke, but I hang out there to find out what's goin on.

Between drags on their cigarettes, everyone is grumbling and bitching about how unfair it all is.

I can feel the wind blowing ripe with discontentment.

The hallway of the main building was crammed with students sitting on the hard cement floor, their books perched awkwardly on their knees as they vainly tried to do their homework.

Some of them were obviously pissed off. Others were staring up into the air, waiting for the day to end so their parents could come pick them up.

A few were pushing through the crowd, loudly commenting on how bad it was, while still others looked out and out disgusted. The less enthusiastic simply drooped like summer's last wilting flower.

As I storm through the hallway, I make a "command" decision. I stride out into that parking lot, past where a few of the good counselors, administrators and teachers are taking a cigarette break. They sheepishly avoid my gaze.

(They remind me of those high school kids who always waited til the last minute to dance, and wished they had started hours before.)

I need a minute to clear my head and think.

This is ridiculous! Look at all these people. None of them are getting what they want, and they're all unhappy. There's got to be something we can do about this! Look at all this anger! Look at the energy we're generating!

I know from bitter personal experience that anger alone doesn't do a whole lot of good. Then it hit me, information is the answer. What I have between my ears has always been my best tool to create true change. That coupled with the fact that apparently I'm just not afraid of any fucking thing.

I started checking it out.

First question: "Of what value are we students to this school?"

I discovered that our school was paid on an attendance basis. Each of our warm bodies represents $2,000 per day for the school. How does that work? Well, it turns out that our scholarship monies aren't paid directly to us.

The government, in their infinite wisdom, knew the students would fritter away the money. So, they set up these trade schools on the basis that if the school offers job training programs to lower income students, the school will directly receive the funds.

What a sweet deal for them! The school gets paid regardless of whether we learn anything, or even if we graduate!

If you've ever wondered why on earth a college would take attendance like the students are all five years old, it's because that's how they get paid, people! That's always been the bottom line for schools, so the truth of the matter is, we are the reason all those administrators and teachers (even the ones with their feet on the desk) are getting paid.

OK, fine.

But, what happens to their little plan if the students don't show up? What if we all decide to walk out and not attend their stinkin no-book, no-desk, no-chair, "I don't give a crap about the students" school?

What will it do to the school if it sits empty, without our $2000.00 butts sitting in those classrooms? What kind of an impact would that have?

All the anger in the world can't match that power.

But what if no one knows about it and they try to shut us up?

I'll make sure that doesn't happen. Tell you what I'm gonna do ... I'll call our local newspaper. I'll tell them there's gonna be a big protest at South Bay College, because the students are getting ripped off! I'll complain that we don't have enough books, or enough desks and that we don't have teachers who'll teach us. Oh yeah, I'm gonna do that. Right now.

At school the next day I marched right down to the cigarette break area of the parking lot, and told all the grumbling, mumbling students exactly what we're gonna do.

"Here's a bunch of construction paper and spray paint. We're going to make signs!" the grumblers immediately wondered what the signs should say.

"You just write down some of the stuff we're all sitting around saying right now! Then we're going to tape them onto these yard sticks that I brought with me. I don't care how sloppy it is. I don't care how beautiful it is.

The media is going to be here on Thursday. Everybody show up with all your righteous fuckin anger, because you know what? We're all gonna walk out of this school! We're not going to attend until they give us what we want! They're gettin our money anyway!" I ended in a huff.

I'll tell you what. There were so many angry people who agreed with me, they started making the signs right there in the parking lot. People were angry. People were sick of dealing with it.

But, most of all, people were just flat out fuckin amazed that there was anything that could be done about it.

Believe me, there's a lot you can do to change things. You need to do some research on the problem, come up with a brilliant plan, and then take action toward that change!

Sure enough, on Thursday, there had to have been at least 50 students there.

I've never even seen 50 students there on one day, but people had told other students about the walkout.

I guess, they'd all been sitting around at home, getting pissed off because they had gotten ripped off, and didn't know quite what to do about it. They had dropped out for the same exact reasons we were protesting.

The students showed up in force, all right. The local media hounds were there as promised, with a photographer busily snapping pictures while we protested.

About half of us had signs, as we stomped around in a circle, shouting.

"We won't take it! This is screwed!" I mean, we were LOUD.

Cars going by were slowing down to listen to us, some even honked their encouragement. That's then we really starting getting animated. It was quite a circus.

And you wanna know what? On Friday morning, the Dean

of all nine campuses of South Bay College came down here to talk to us. He wanted to know what he could do to end the walk-out.

(Turned out I was right about the money!)

A bunch of people pointed me out as the instigator of the walk-out. I grabbed a couple of other people who were instrumental in making the protest signs, and we all went to the Admin building to talk to him.

We calmly sat down at the table and explained the situation. He seemed mildly surprised that we were presentable, and quite clear, about what had been going on.

"No one will show up again in class until we have enough books for every student who's enrolled in every class. Until we have enough desks and chairs for every student enrolled and, until you get rid of the teachers who think it's their job to laugh at us and put their feet up on the desk and read their newspaper during class." I ticked off our demands on my fingers.

He seemed genuinely taken aback at that last statement, but kept listening as Linda continued.

"You're paying those teachers to teach us, and they sit back and make fun of us. Man, you're wasting our money, and your time!"

We were true to our word. Almost everybody stayed out of class the next week. Most of the students voluntarily joined the walk out, but there were a few hold outs who still attended class.

Problem with that is, most of the teachers have done their own version of a walk-out, in support of the students. Oh, they don't say that they won't teach classes, they come to work every day, but they mainly sat in on the student/administration talks, and don't do much else.

By Thursday of the next week, the Dean returned to the Baldwin Park campus to personally deliver a huge shipment of books, desks, and chairs.

He also fired two teachers, after interviewing other teachers and staff about our accusations. Everyone they spoke to concurred that the teachers in question weren't doing their job.

Although I had learned this valuable lesson earlier in high school, I relearned once again that sitting around and complaining about problems, or just thinking about them and merely planning things, isn't enough to change them.

That's a good beginning. But, until you take action, you're never gonna change anything. I also learned not to be afraid to question authority!

You can fight city hall and win, if you have the right information, unstoppable determination, and are absolutely fearless.

* * *

Cami monitored her viewership's RQ as did any successful 'grammer, but she took an extra precautionary measure in her data logs. She traced each RQ input to the source, then locked onto their outgoing 'gramming requests, so she'd know every string of data they had ever purchased.

She had already picked up a curious new flashwave pattern on the Vidnet, and had this funny feeling it may have to do with her newfound friend (who called himself "Frank".)

Evidently two BayBays had intercepted the information that a Retainee in Rehabco, Terrax had mysteriously disappeared.

(How had they found out so quickly?)

Cami was constantly being surprised by the resourcefulness shown by these Terraxians. Apparently, these BayBays had been tracking the buzz on the Sextel Network and had discovered that the guards had inadvertently "misplaced" someone, that someone being the notorious Milliner!!

The BeenThereDoneThat Blog had posted that the murder had been committed by a lone ranger, an anti-vigilante who was trying to rid Terrax of the most violent of the UVPs.

Cami wondered if, in fact, her new friend had ended up in Retainage. A newbie DownTowner may get lost and lose his cubby, so it made sense he might have ended up in Retention. But, Cami couldn't figure out how in the world Frank could have pulled off a murder while incarcerated in the planet's highest security penal facility.

Still, since stranger things had already happened to Frank in his short life, it was just possible to believe it could be true.

Cami's brows knit as she noticed another pattern of interest leading from M. Dumjon's percomp, it was tracing Frank's every move. What interest did the Guardian have in Frank, now that he was a free PHD?

Was it a personal matter? Cami shook her head as if to say 'No'. It seemed unlikely that a Guardian would find a special interest in one their charges, since their chief aim is "to attain pure knowledge as garnered through masses of information, not specific datum."

His actions were most likely at the behest of his bosses, the MOBsters who controlled the Rehabcos, among other things.

Knowing that she didn't need to check in for awhile, since she had just completed enough programming with the "City Hall" segment, Cami decided to tail Malmo on his next excursion. There was no law against being in the same place as a Guardian, so long as you don't interfere with their official duties.

Feeling like a dark angel, Cami slid into her special skinsuit, the one equipped with a nearly undetectable personal disteldt shield, the only one of its kind.

She would follow Malmo, who was following Frank. Oh, it promised to be a proper shell game to see who would spot who first.

Cami took the connector lines to the Rehabco exit portal and settled down to wait for her prey to emerge from the depths.

* * *

Malmo Plays the Fool

At the end of Malmo's debate, he couldn't help but notice that his RQs were going through the ceiling. He felt warm inside thinking that he'd caused the sensation that was going through the Vidnet right now.

He admittedly had capped on another sad story, but that's the way PR is- whatever is hot, hot, hot, topical, current, and in the now – that's what news is!

If you can tie in your message to a current event, well, you just can't buy that kind of publicity.

The thing is, Malmo had prepared this debate annuations ago regarding "the need for violence in the human genome". At that time he'd actually put a lot of thought into it, had based it on countless studies of ancient history.

Not only had he accessed the TDM (Thermal Disimaging Machine), which had information from people who had lived before and all their prior knowledge, he'd synthesized the information over many annuations of being a Guardian by observing his Rehabco Retainees behavior's and thought patterns. It was still amazing to Malmo that even though the Retainees knew they only had one annuation to wait until their sentencing, some of them resisted disciplinary measures. Then they would be further restricted in their access to decent jobs.

Why did they resist when they knew that there was no way out? These incorrigibles knew that resistance was entirely futile, yet a few still resisted their work assignments, once out of Retainage.

These miscreants escaped and chose to live on the OUTside of the Hives, to become the lowest of the low, a SadSack. Some of them survive for a short while, but not many.

Malmo had noticed a disturbing trend in the profiles of the most rebellious of these PHDs. They happened to be the incarcerated UVPs, the MOB's own cast off children. In order to trace their gen-ennge lineage, he accessed personal datum streams from families who owed him chedits, leaping recklessly above his own security level.

What he discovered could make him a rich man, one whom the MOB would owe a huge favor!

Even though the MOB was his employer, Malmo now knew that there was a connection between the MOB's and the Senselle's Gen-ennges. And the viewership was being lead to believe that the two parties were enemies, each espousing an entirely different set of ethics.

Faugh.

It simply isn't so. It appears that the MOB and the Senselles have conspired to control the viewership through back room politics. To Malmo, there was something inherently wrong with this manipulation of the masses.

He agreed that violence should be removed from the mainstream of the population, because there is no room for random violence on Terrax. Everything has environmental restrictions, laws, rules and regulations, due to the sheer numbers, now at 26 billion PHDs.

There must be complete order or there'll be utter chaos; a titanium meltdown, the destruction on a massive level that even the ancient worlds could not have known.

Regrettably late, he realized his bosses had conspired with the Senselles, their sworn enemies, to restrict the flow of violent genes into a very few PHDs (thus the UVPs were created).

They then agree to allow one UVP at a time go out and create controlled chaos among the viewership.

He didn't divulge this information during his public debate, because after all, the MOB is his employer. They'd cut off his chedits, or worse.

Feckit, let the viewership be finagled out of their chedits. He knew what was what, and who paid the bills.

After finishing the debate, he knew he'd done a good job, but he didn't feel that great about himself. He never really did. He always considered himself a rather mediocre person, and so he was.

Like many other "armchair intellectuals" he could read the works of others and quote them accurately and at length. He attributed great quotations to the correct authors, but he rarely synthesized the concepts into his own original thinking.

He simply regurgitated the information back to people and threw so many references at them that they thought that he was very, very, smart.

Malmo was determined to change all that with the results of a new experimental study he was conducting.

During the last Wheel of the Year, he chose one of the Retainees at random, and decided to make his life a living OUTing.

I enjoy studying how PHDs react under imposed conditions of extreme mental tyranny for an annuation, then I publish a study on the results by Soyal.

It that will make me famous.

Then, let them snigger!

I chose Frank as my first test subject because the instant that Frank came into Retainage, I took an instant disliking to him.

There's something wrong with this gemmat. I've seen all sorts of genetically engineered people, and there's something....I dunno.... unnatural about this Frank character.

Frank reminds me of one person one nano, then a completely different person the next. It puts me off my oats. And I sure don't like that they made Frank a Trustee and let him wander around unescorted. It gives me the willies.

Since I'm not able to really sleep properly, I take about seven or eight little cat naps during my work cycle, but for only a few minutes at time. While awake, I endlessly review Frank's data tapes.

Not only the physical data, but also the mental biofeedback loops coming in from all my Retainees, just for comparison. This is something that few of the other Guardians do.

They don't pay much attention to me and my insignificant study. They watch their own numbers. If their Retainee's biomeds fluctuate enough, their Guardian may finally take heed and conduct a one-on-one webinar with them.

This kind of one-on-one is designed to keep the little lambs in place until they get their next work assignment and go happily on their way.

I've got a good job, or so the MOB loves to tell me. It's true. Even if one can't stand a Retainee in their care, at least we Guardians don't have to put up with them for more than one annuation.

I see my job differently. I know when one of these little

weasels is going to go AGGIE on us, and I just can't let them go without a full background check.

When I spoke about UVPs during the set-up of my second premise in the debate, I mentioned "The Milliner" as a primo example of gen-ennge perfection in the concentration of Ultra Violent genome engineering in a PHD.

I noticed some interesting reactions in Frank. Oh yeah, I've been monitoring Frank/ie Boy very closely. Of late I ordered nano-by-nano reports on him, including fluctuations in mental data feeds, biomed readouts, energy signatures and audiovisual surveillance.

I've spent endless cycles observing the local Vidnet, which has taken up a huge percentage of my personal accessing allotment. That may seem a bit suspect, but as a Guardian I can do pretty much anything I want with my chedits.

Except flaunt the MOB, of course.

At any rate, after talking about The Milliner I notice Frank is starting to fluctuate between Frank's two personalities, whom I've had dubbed Fran and Frank.

Fran is the softer, more sensitive side of Frank. When Frank is Fran, s/he changes moods quicker. S/he goes from highs to lows rapidly, and for no apparent reason.

Actually, I hadn't really noticed Fran (the second personality) until about three quarters of the way into Frank's retention period.

When Frank is Frank, he pretty much smiles all the time (but it's a kind of smile that says he's laughing at me, not with me). In fact that smile is kinda creepy. It's that stupid, crooked grin that drives me completely gonzo. I wanna wipe it off his pale, thin face once and for all. He's so smug.

It really blows to know the punk will get off with such an easy work assignment, but since he's in here for some petty bullshit offense, I guess the sentence does fit the crime.

Yosaka pulled some strings with Grandpa's peeps and got Frank assigned to the Calava Corp upon release, which is one of the cushiest jobs in the Big Four Industries. It makes me wonder why the twerp is so special, that he'll be spared going to the Aerosuit Industries, where the MOB can keep a closer eye on him.

Too bad. Working for the Aerosuit Industries is one of the low-down dirtiest, nastiest, smelliest jobs in the Big Four, where they make skinsuits, and their tonier cousin, the Aerosuit.

It stinks abominably; a truly wretched job usually not assigned to minor white collar petty crimes, like the one Frank is in for. He was caught "pinching" the account of a minor MOB employee (a DownTowner).

He hadn't taken much when he was caught; a few snippets of a chedit here and there, not really enough to be that much of an offense, but enough to get him slapped into Retainage for an annuation.

That's one of the things I still don't understand. Since Frank had worked for the MOB when he committed the chedit theft, why wasn't the MOB taking care of this internally? Why wasn't Frank simply demoted or devolved, rather than actually stuck in Retainage for something this minor?

For some reason, the MOB had felt they needed to tuck Frank away for a short period of time, and it wasn't until he was in his eighth Pana, that I began to suspect gen-ennge involvement.

When Frank began fluctuating more rapidly between Fran and Frank, he appears to be remaining Fran for longer periods of time.

It also seems like it's happening against Frank's will, and that he's having a harder time trying to hide the changes. Frank appeared to be physically changing as well.

When he's Fran, he looks softer- somehow- like his bones are becoming more rounded. He doesn't look so angular. He starts putting on a little weight around his hips and thighs.

Malmo didn't know what the significance of all this was, but he remained watchful. As the 11th Pana of Frank's waiting period rolled around, just a few cycles before Frank's work re-assignment, The Milliner was mysteriously killed.

Within hours, Frank's cycles began to fluctuate madly back

and forth.

Then all of a sudden, the day-cycle after the inexplicable death of the most infamous UVP on the planet, the old Frank was back – with that irritatingly crooked smile.

It's...strange, odd, discomfiting.

Dammit, I really hate feeling this way.

Lucky for me, I just got a peep from upstairs. Seems they want me to tail the cocky bastid to his next work assignment. They want to find out if there's an AGGIE connection from Frank to Wad Medani's MMs, or the Senselles. Either way, I could care less, I just want to dig the dirt on Frank/ie boy and be there when he's put back in Retainage.

Then, he'll be all mine.

* * *

Frank/ie is Free!

The change happened right in the nick of time.

Frank/ie had been released from Rehabco without incident. Her next work assignment was a cush job with the Calava Corp., which was beginning in three day cycles.

She almost couldn't believe that she escaped without one of The Guardians realizing that she had killed The Milliner. And, she'd managed to walk away from Rehabco without them catching her own personal freak show.

Sometimes blind assumptions can work in your favor. The fact that Rehabco was the highest security level facility available, coupled with their tightly structured lives led the PHDs to believe that no one could escape from Rehabco.

Yet the same blind faith that allows them to hold diametrically opposite ideas in their minds at the same time, told them that no one could be killed without their knowledge. This psychic dissonance caused them to come up with the only "answer" their minds could accept, that some super person had somehow mysteriously killed "The Milliner", then escaped.

Frank/ie was starting to go through the change that would

return her from being a gemmat, back to her former gender as a gemma. She mustn't get caught publicly going through the change or she'd never escape the rigidly cold world of Retainage.

If her secret were discovered, the next time she may end up reassigned to the off world Rehabco on Phaeron!

Her check-out process seemed interminable, but finally, the Rehab ID strip was imprinted in her left wrist. Frank/ie rubbed the place where the implant was, feeling like she'd been branded.

She was looking forward to her free time, although she didn't have a clue what she'd do with all that time to herself.

Unaware that s/he was being discreetly followed by her nemesis, the Guardian Malmo, s/he took off for the nearest underground, WC sponsored hostel.

Malmo initially followed her with remote operated bots, rather than physically. He wouldn't make his move until he had the evidence that Frank had committed a crime.

Malmo fervently hoped it would be something worthy of stowing Frank away again.

* * *

Frank/ie knew a lot more OTEC secrets than Auntie Vee or Wad Medani were aware of. She'd been absorbing information, much like a baby in a womb who's subconsciously absorbing the meaning of spoken language, interpreting the emotions and actions of the people outside its birth chamber.

Frank/ie remembered catching bits and pieces of information about the outcasts of this particular society, who called themselves the AGGIES (the Above Grounders). The rest of the PHDs on Terrax referred to them as Sadsacks, because they'd taken to living in the organic sacks clinging to the OUTside of the Hives.

They inhabit the sacks, which are similar to a fungal skin growth, for only a few cycles at a time. The AGGIES use this natural material to fashion crude, tent-like structures in which to live. The Sadsacks, as they are known, are only temporarily attached to the OUTside of the Hives, because the harvesting machinery must scour the Hives regularly. The Harvesters clean off the biological incrustations in order that the Hives continue to generate enough electricity to supply the entire planet's machinery, including life support systems for the Hives and Warrens.

This necessitates a highly coordinated lifestyle. The AGGIES live in constant motion, outracing the harvesters, jumping from bubble to bubble. It's a nomadic life, one filled with the strictest of discipline.

The Harvesters come around only every so many day or night cycles, they're on this bizarre rotating schedule. You only have so much time to stay in a particular place, then you have to move on.

Information is key, because you have to know exactly where each of the machines will be working, in order to stay safe for the next few cycles. Quite often, AGGIES are caught unawares, and killed. Changes in scheduling due to repairs on the machinery can interrupt the rotation of the cycles, and there can be unscheduled harvestings.

Being an AGGIE rebel often means dying ingloriously and alone.

The WC has a two-fold purpose in cleaning off the Sadsacks. One is to ensure that the power generation is clean and pure, and that we have enough to supply the whole planet, but the other is more political.

The AGGIES are considered outcasts of society, and dispensable.

Furthermore, the power triad is beginning to consider them rather problematic. The WC and the MOB (the old power regime) don't appreciate the AGGIES giving people weird ideas (like being able to live OUTside). The attempts to breach the building's security were causing more toxic leakages to occur.

Whenever the AGGIEs decide to breach the enviro-barrier, whether by coming up from the Warrens to the OUTside or by going OUTside from within one of the Hives, the enviro toxin exposure to the life support systems becomes critically high.

As the Hives are designed to be completely self-contained systems, causing environmental leakage within the buildings is a very serious offense on Terrax. It's considered the highest nonviolent offense one can incur, often punishable by permanent banishment to Rehabco, Phaeron.

While she was being reconstructed in Auntie Vee's Uberlabs, Frank/ie had inadvertently absorbed a lot of AGGIE information due to tiny transmission leaks from Auntie Vee's percomp into her electro-matrix.

She knew that the WC and the MOB agreed that the Sadsacks were becoming more and more of a problem. They knew that the AGGIEs were becoming more aggressive in their beliefs, proselytizing to the common viewership, and spreading their beliefs to the precious Enbee viewership as well.

Support for the movement of aboveground, unmonitored living was seeping into the media, which neither the WC nor the MOB were too happy with. Evidently, the AGGIEs have a powerful connection into the communication system, and an unauthorized access jack-in which they use to broadcast their beliefs far and wide.

Still, Frank/ie had to admit that these AGGIES seemed to be the only PHDs capable of outfoxing the WC, and even the MOB, with their unbendable willpower.

Frank/ie definitely needed to meet up with these people. But how to find them?

She had hidden away in her free WC sponsored overnight hostel for the first day-cycle, too exhausted and confused to do much else. The let down from having killed the Milliner seemed almost anti-climactic.

Then the walls started to close in around her, and she decided to explore the Warrens for the two day and night cycles she had left to herself before reporting to her new work assignment.

She wasn't looking forward to her new work assignment, having gotten a sneak peek when she'd accidentally jacked-in to a BayBay who'd 'grammed his work cycle at the Calava Corp. She thought she'd die of boredom in those few nanos, she couldn't imagine what it would be like working there shift after shift, cycle after cycle, for the rest of her short life.

Feeling adventurous, Frank/ie decided to check out an underground bazaar she'd just seen an advert about, going on somewhere near here. Leaving her hostel, she boarded a public slidewalk going Downtown.

(She had to be super careful for the next couple of cycles, to make sure that nobody followed her. She wanted to return to this same hostel with some assurance it was safe, at least for tonight's cycle.)

Rumor on the Sextel secured Channel had it that the AGGIES had secret underground meetings. She vaguely remembered that the FLEE Market had something to do with it. She knew she should equip herself for the excursion, but hadn't the slightest idea of what she'd need.

Going on instinct and pure adrenalin, she entered the underground bazaar and intuitively started strolling down a row of munchie shops.

She had just enough chedits to buy a sweetshot (the equivalent of a protshot, but rather than shooting it for subsistence; it was dessert.)

These dishes had a heady little rush, and came in different flavors. The shops offered free samples of the most tempting, luscious, dainty morsels she'd ever imagined.

The samples didn't satisfy your craving, but rather, enticed one to buy one more, then one more, then just one more taste. The whole point was to purchase enough of these morsels in a row, so that you could taste the full flavor of one exotic dessert.

Thoroughly enjoying herself, she lost track of time, strolling through the brightly decorated shops, tasting a bit of this and a byte of that.

She was brought sharply back to the present, when she heard a very quiet murmur from two people who were scurrying by. She registered this as odd, then realized why with a shock.

She had not seen *two people together* going anywhere in all the underground. She'd seen individuals, people with personal force shields around them, similar to disteldt shields, but she had NEVER seen two people walking together.

She thought it curious enough that she decided to follow these two for a bit.

Unbeknownst to her, one of them was Arpad Reitan, who was on his way to an AGGIE meeting. He was curious to learn more about the technology that they used to survive in the OUTside elements.

The conditions the AGGIEs had learned to master were similar to those he must simulate within his own deep-space going craft, which he would use in the pursuit of his dream of becoming a HORSE.

He happened to be accompanied this evening by a surprisingly familiar face that Frank/ie had seen around the Uberlabs many a time. It was the Odissi temple dancer, Damaria Rotiarte.

Frank/ie wondered what in all the World's a top ranked Senselle would be doing with a lowly OTEC techie. She knew that Arpad worked for both for the MOB and for OTEC, which is an enigma in itself. Obviously, he must also have connections with the underground, too.

Who the heck was this guy? On impulse, Frank/ie decided to see where they were going.

Frank/ie didn't know much about her future, but she did know she wanted two things for sure.

Never having been to the FLEE market before, she had fancied a taste of the deli-munchies sold in the kiosks along the way.

Check.

Next, she desperately wanted to know more about this Arpad Reitan.

She had first seen Arpad while she was in recovery in Auntie Vee's Uberlabs. She had noticed him when he came in to report the progress of an OTEC assignment, and she'd liked his quirkiness, his voice and his stride.

Everything about him was exotic and enticing, completely different from her original meshling, Teekum Dowdy.

The thing is, Arpad had paid absolutely no attention to her while she was in the tank, and had no idea that she could morph from one gender to the other.

Who would have thunk it?

Now she was feeling that itch to have sex again.

She knew she had been oversexed when she was with Teekum, although he never really quite appreciated it. They would jack-in as much as they could afford to, but it was never completely satisfying for her.

She knew she and Teekum couldn't afford the Senselle experience, but once she knew what that was like, she'd never be satisfied with simple VR 'gramming. She was becoming more and more restless, like she just couldn't settle into her new life.

Life as an UpTowner, with her comfortably lazy meshling, had been only partially satisfying her tastes. Now that she'd seen this crazy flamingo of a man, all she could think about was him, the wild Arpad Reitan.

She'd overheard Arpad whispering something to Gramps, in a three-way peep with Wad Medani, that they intended to place an agent inside the AGGIE movement. They wanted to find out if the MOB was connected to the AGGIES, and if they were assisting them with the mysterious ROV transmissions.

Other questions begged answers, like whether the MOB had procreated UVPs for their own purposes, and how to stop them.

Thus far, the OTEC Gen-ennges suspected that the MOB was in control of the UVPs, but they weren't worried, yet. They had followed the rise and demise of The Milliner with curiosity, but not fear.

Personally, Frank/ie could think of no worse evil than a PHD who had been specifically engineered to wreak hideous violence on innocent BayBays.

She knew that OTEC could easily dispatch demons like The Milliner, anytime they pleased. Instead, they were more concerned with a stupid ROV, some machine that had gone crazy, damaging the Ocean Dome in the southern hemisphere of the Atlantean coast. Some drivel about it sending unauthorized communications off-planet.

Who the VEE cares about that when her family had been

nightmarishly torn from her, and her own body ravaged, left to die?

Frank/ie didn't catch all of the convo, but she also discovered that Arpad was a HORSE-in-training. His job was to wrangle the biggest ice comets and planetoids for recycled water, metals and raw material.

He became involved with OTEC because the Senselles needed to hire his HORSE services completely off the radar. It would be his job to infiltrate the AGGIES, and attain the materials and connections necessary to launch a solo mission.

For his part, Arpad was determined to become a HORSE. (Little did Frank/ie know that he had been banned from that particular work assignment level by the WC, because of his extreme height.)

He'd chosen to break into it anyway.

While the three men spoke earnestly about the underlying inferences of the ROV accident, her eyes remained on Arpad, noticing that he was lean, tall and rangy. With that spiky hair, he almost looks like an ostrich, she thought.

Still, she felt strangely stirred to touch him, but realized it would be too dangerous, for both of them.

Testing her newly acquired shadowing skills, she surreptitiously followed Arpad and Damaria as they disappeared into a darkened archway, leading to defunct offices within a nondescript building.

She was sure they were going to a forbidden place, by the way Arpad and Damaria quickly checked around to see if they were being watched.

As they ducked into a small room, Frank/ie took her first step into official rebellion.

An AGGIE meeting!

I'm thrilled and nervous at the same time. I've been warned that I should never attend one of these meetings, and I can't ever let Gramps know what I'm doing. I don't think he'd appreciate it, or approve. I know he wants me to stay low-key, but I don't think any of 'em have one freak'n right to tell me a thing!

I have to find a way to see Arpad again. I wanna know what the Vee he's doing with that temple slut Damaria.

As far as I know, the Senselles have no connections to the AGGIES or to the demon MOB. (I suspect the MOB had something to do with the UVP who destroyed my life.)

I don't know how The Milliner found us or what purpose it served to allow him to do what he did. I know everything else on this damn planet is planned out to the nano, so why shouldn't I think that was too?

When I find out who sent that creature – they'll pay. I don't care how big my target is or even, Vee help me, who it is. I'll go to the very top of the food chain to get my ultimate vengeance. All will come to be, in time.

For now, I've another, more pressing need.

Frank/ie had a burning desire to evade her next work assignment, and the only way out of assigned work is to buck the system entirely. The crushing boredom promised by her next assignment was enough to make any PHD go OUT of their mind.

Originally, in her previous life, she had been an OTEC techie, but she could never go back there. Once she morphed back into her former self, people would recognize her and start asking awkward questions. She was, after all, "officially" dead.

But since she'd never worked for any of the Big Four, and since the Calava Corp. was the least MOB infested, that's where Gramps wanted her to work after she got out of Rehabco.

The Calava Corp., where they produced protshots, was one of the stinkiest, most repetitive work assignments that a small time Retainee could land. But it would be completely safe.

Remembering the 'gram she'd seen of the work, Frank/ie knew it would bore her to death. All the workers did for an entire cycle was to monitor the huge machines that recycled the raw materials – whatever it was – that ended up becoming the protshots in their various flavors. Most of the raw material was funneled directly into the system that distributed the protshots to the cubicles.

There was no direct interaction between the employees or

even the machines; one simply watched a computer screen.

She was told it was important work because without their daily protshots, all PHDs would die within days.

At this point, to Frank/ie, dying seemed like a mercy!

That was another reason she was considering joining the AGGIES.

On a practical level, they may be the only group of PHDs on the planet who can accept me for who I really am. The whole truth about me.

Not to mention the extra perk of being close to Arpad. I'm being overtaken by this intense sexuality kicking up inside me. I fiercely need my chance to talk to Arpad, to get some time alone with him.

I remember seeing him one day (while I was still in the tank at Auntie Vee's Uberlabs), just as I was beginning to wake up.

I started to become painfully aroused. Unfortunately, I hadn't noticed I'd morphed into a man. I felt an unfamiliar slap of flesh against my thigh, looked down and to my utter shock.... found I had a penis.

I was stunned.

"By Vee, how did this happen? I have a dick, and I want a guy! Ugh!"

You can't imagine how confused I was! It's like I would think about him and get hard! It's ... oh my God... not what you expect after having originated as the female of the species.

I must say, there's definitely a drawback in having to live as only one or the other gender, instead of being androgynous. The Andros have it all! They can be whatever they want to be at the moment and not really worry about limitations, stereotypes or characteristics. They don't need to limit themselves, or worry whether they're doing what they're supposed to do or what is expected.

It's all such a bunch of crap! It's enough to make me want to go OUT!

Anyway, once I got out of Rehabco, I changed back into a female. And that wasn't as easy as I thought it would be, either.

You see, before "that day my life changed" I was much more comfortable in my body. I seem(ed) to be in alignment; I knew exactly what I was going to do, how to fit in. I knew what I should look like, and how to portray myself.

Now I have to learn to be a woman all over again, knowing how men really feel about us.

I guess someday I'll get used to my secret self. There are subtle differences and obvious ones.

Mainly, Frank is a bit rough and stand-offish. He wears a snoid grin on his face, like he's saying "I'll never really give you my real self. I don't talk a lot, but listen to people inside my head, offering critical commentary as I whistle a favorite tune, "Grist for the Mill."

Even though physically I'm only one gender at a time, mentally I'm always both Frank/ie and Frank.

I feel completely separated from the rest of the world. A freak without a show.

It's only me, myself and I.

* * *

Coming to, back from her mental wanderings, Frank/ie found the meet-up place and ducked into a tiny cubby. Without hesitation she jacked-in to the AGGIE meeting mindgroup and prepared to meet Arpad in the full splendor of 100% VR 'gramming.

Barrage rides the light fantastic

Barrage concluded the peculiar webinar with the great Asna, feeling less rather than more informed than before the call. As usual, Barrage displayed his dose of cynical, snide, completely-in-control 'tude while Asna telepathed his most pertinent thoughts, immediate wishes, and end results regarding Barrage's next assignment.

This day-cycle, Asna appeared to be agitated, although showing no outward signs. Barrage did note the lack of visual image accompanying the usual spate of information and data download. That was unusual.

However, Barrage was in no mood for antics of any type, and therefore appreciated less drama and more hard facts, more misdirection, and less diddling about.

As Barrage contemplated the information streams, he realized that Asna and Wad Medani had just had their first face to face meeting in over a century. (A century was many, many, many hundreds of annuations, which is much more time than the average PHD could possibly comprehend.)

Barrage gathered that their meeting had been as histrionic as it had been historic, both indisputable leaders claiming that the other was lying.

This was probably true in the literal sense, but not in the most important sense of the word, which is: they were both completely innocent of how this ROV had been programmed to transmit communication signals to a satellite located within Phaeron's orbit.

Barrage had been informed that one of the Rehabco satellites was directly in the line-of-sight of a most mysterious data transfer. The satellite hadn't received the information, this much Asna had been able to ascertain from one of his space cowboy buddies stationed out there in the wild yonder.

Upon notification by his own sources of the illegal transmission, Asna's commercial fleet had immediately been dispatched to Phaeron, to make a physical inspection of the satellite there.

The HORSE on call was Talon Gaet (a nestling sib of the famous bard of Terrax, Phoenix Gaet). Talon was a loyal submember of the World's Council, and as an absolutely steadfast employee of the MOB, his work was unquestioned.

His only interest was in racking up as many chedits possible by looping in the juiciest, richest planetoids, comets, and asteroids, for raw materials and hydrogen. The HORSEs called it "hydro heaven" when they scored a chunk of rock containing water.

Talon's recon team found that the satellite system wasn't

breached, the tower receiving the transmissions from Terrax showed absolutely no signs of tampering. The equipment hadn't been touched since its last scheduled maintenance.

His ship, the Sodomy Bear, silently cruised by the Rehabco, Phaeron facility, but they reported that everything was normal there. The outgoing transmissions indicated no signs of trouble from either its automated staff or the inhabitants of the Rehabco.

Talon reported in to the MOB motherboard. They wanted him to download the last annuation's data records from Phaeron. The whole shamu, all data feeds from the residents to the automated servo-bots who attended them!

Vee, that was a lot of bytes! It was going to take eons, Talon sighed inwardly to himself.

In the meantime, his crew had been instructed to physically inspect all the equipment piece by piece, but again they found nothing out of the ordinary. They duly reported that all the life support systems checked out, as did the satellites orbiting Phaeron, both in receiving and transmittal modes.

All data streams matched the known data records at the time. There was absolutely nothing untoward.

So, Barrage wondered, what had happened to the ROV? Who had reprogrammed it? Why it was transmitting to Phaeron? Neither Wad Medani nor Asna knew.

But, they suspected that the other did know more than he was telling, so each held back some vital piece of information. Basically, they had lied to the other. In keeping back information instead of sharing it they had unwittingly exacerbated the situation.

They'd each left their meeting singing another aria of the insane opera that we call "life" here on Terrax.

Of course, Asna wasted no time in laying the problem on me. He figures I'm never doing anything of import, just sitting around waiting to be his clean up man for the MOB's dirty messes.

Let him believe that, for now. It suits my purposes.

As for really changing the world around us, I don't hold him up as my role model.

It may surprise you to know that my role model is Cami Chameleon. She could teach Asna a thing or two about real change.

I figure she's one of the only Senselles left who can still surprise me. All the rest of them are too jaded, too obvious, or too one dimensional. They bore me. But Cami – she keeps evolving from one persona to another in her adventures.

So I tune in.

I know, I know, admitting to watching them is tantamount to admitting to a guilty pleasure. It's something that I should probably be terribly embarrassed about, but I enjoy watching her too much to let that bother me.

Although I'd love to get to know her, I know I'll never meet her. We travel in different circles, and hers is a bit too rarified for my likes. Additionally, she's on the opposing team to mine in the current political battles.

But, I keep getting sidetracked from my original report. (I guess I'm not taking my current assignment seriously enough.)

Perhaps I should consider putting a team together for this assignment.

Of course, that's the one thing I've been instructed not to do. I'm supposed to accomplish this work assignment entirely on the sly. Meaning using my own funds, and no help.

The problem is, I'm already looking at several people to help me with it, trusted "players" from my own team.

One of them is Malmo Dumjon. He's perfect because he already belongs to me (he works for the MOB), and because he's corruptible. He has egotistical desires, which means he has weaknesses I can manipulate.

Asna wants me to tail Malmo, who has been assigned to follow Frank/ie. (I can't imagine of what interest that person can possibly be. She's some nobody who just got out of Rehabco and was assigned work in the Calava Corp. Small potatoes.)

As far as I'm concerned, it is Arpad (a wannabe HORSE, and a real freak of nature) who is the more interesting prospect.

The story goes, he's too tall to be able to fit in a moplane, so he can't get transport to a spaceship. Even the mini-stealthpods aren't big enough to take him up to an outbound ship. What a loss to the program.

For some odd reason dear Asna wants these two nobodies followed into an AGGIE meeting. Apparently, the MOB needs to know more about the AGGIES.

What I know from the intel I have, is that the AGGIES don't have a lot of chedits. They're not an economic strength. They have no particular ins or ties to the current power triad. They're a gaggle of fools, really, lead by the troubadour poet-fool, Pavarti.

As if anyone would want to follow a leader who's lived in a SadSack, those nasty hollowed out pods of biofouling growth clinging to the OUTside of the Hives.

These numbnuts go OUTside with only as much enviro-protection as their poverty can garner.

They do all this to prove it can be done, to avoid the Vidnet, and to avoid being identified.

Yay team.

By doing this they also "avoid" a WC issued cubby, replete with prot-shots, clean air and water, and at least a modicum of media entertainment.

By Vee, PHDs have to do only the very minimum to survive, and this is what the AGGIEs are resisting with such vigor?

They can earn an assignment in one of the Big Four and work their way up the social cell structure, or, if they can produce original 'gramming, the sky is the limit.

After all, original entertainment is the only thing that's worth producing.

As for yours truly, my true end-game is to find out who programmed that ROV.

Right.

The MOB expects me to do what all of Wad Medani's MagicoMathematicians can't do. Asna wants me to figure out a puzzle that neither the unbending Wad Medani nor "the great Asna J." can solve.

I think they're not paying me enough for this.

And now Asna thinks it will help to have me trail around vidiots like little Frank/ie, a loser like Arpad, and the worker-bee Guardian, Malmo. Historian or no, he's pedantic. He's sophomoric.

Ah well, to the dance!

First step, put on my masquerade costume and attend an AGGIE meeting.

Looks like the only fun I'm going to have on this assignment will be how I get there.

Should I ride the energy connector waves or do I hop on the slidewalks like a mere pedestrian?

I say, leave behind the workaday, day-cycle, rag tag and go high class on the connector lines of pure energy.

I call it "castenading."

4

The AGGIE Meeting

Welcome to the Jungle! the old style neon-gas sign intermittently glowed a dull orange.

So, this is the cyber punk scene, Frankie thought to herself.

Once inside the AGGIE meeting, she was glad she had morphed back into her original female form.

No one knows that I'm the PHD everyone's talking about – I'm The Anti-Vigilante. They even have a nickname for me now; they call me "AV". I'm not exactly sure what they mean by that, but I'm hearing it again and again on the open interchange jack-ins using the local motherboards.

The peeps think I killed the bastid for "the good of the general viewership".

They couldn't be more off-base.

I can't believe the full on max VR activity here, even in this dive. Their building is krunk (meaning foul, filthy, and disgusting), but its security is hard wired. We can all access the lowest levels of free WC Vidnet at the same time, allowing us to jack-in simultaneously on the local secured board, with no detection from the WC free-radical killbots.

It's as close to physical connection as is possible here on Terrax.

Even we "rebels" aren't crazy enough to try and physically touch each other.

I'd shadowed Arpad to see where his jack-in portal was located.

Evidently, he's known around here. I think he's connected to their leader, Pavarti, in an underground sort of way.

During the intro's, a brief disruption in the local energy grid signaled an attempted breach of our firewall, but Pavarti had a mini-program ready that devoured it, unnoticed.

After that first disruption with the communication system, the rest of the meeting flowed flawlessly. Pavarti set us all at ease with that first demonstration of his firewall of safety. As the social introductions ended, the political portion of the MeetUp began.

I'm not that interested in this part so I'm going to do "double duty" and listen while I 'gramme a few off-topic thoughts. Something inside of me is changing and I want to chronicle it.

I'm not sure which direction I'm going, or how much of myself I might lose along the way. But, what good is my old self to me anyway?

How can I continue to dwell in the past, living and thinking like I did when I was the sweet, innocent Frankie who thought her little problems with Teekum Dowdy were the beginning and end of the world?

For a moment she let her thoughts turn to Teekum, about how he was always leaving the portal door open for too long, which he thought was pretty funny. A light- weight problem. To Teekum, everything had been a joke.

Now, we all know different.

Because of his silly game, I was raped and tortured, then thrown OUT. Although I was "retrieved" and have recuperated, it's not much of a life. I've been given an assignment as a drudge, working for Calava Corp. for the rest of eternity.

A DownTowner! This is something I never thought I'd sink to, because I was a superb media producer.

I'd been employed by OTEC, for Vee's sake, one of the biggest megalocorps in the known worlds. Dubbed an "original" thinker, I'd created 'gramming that maintained the powerful machinery that drives this world.

Now, they devolved me down to working at the malodorous Calava Corp., reduced to the lowest rung on the ladder. I'm desperate to get out of that work level assignment, and seemingly all alone in my attempts at doing so.

Devoid of anything resembling love in my life, now I've become mentally attached to Arpad Reitan.

I'm entranced with the idea of being entranced with him, if you will.

Isn't it always the case? When your heart is breaking, don't you look to the nearest available fool and say to yourself, "Oh! Someone else I can latch onto."

And he's as different from Teekum as day cycle from night. Arpad is a wild cannon, shooting around the Cosmos. All he wants to do is get away from this place, and you know, that idea's beginning to sound better and better!

What is there here for me?

I'm a freakazoid. I have absolutely no desire to be who I am, because I'm completely alone and one of a kind. I morph from one impossibly shallow gender to the other – flip flopping between extreme stereotypes. If only PHDs would realize what peace there could be in androgyny.

But I'm not afforded even that much peace. I'm either physically one or the other.

And, with the intense emotional release after disposing of The Milliner, I feel emptied out. I don't know what to do about my future.

I just can't go back to working for one the Big Four, like a normal fellow, because with me switching identities, I'll eventually catch the attention of that krunkin Guardian, Malmo.

He'll follow me around until he can get me back into Rehabco to have his way with me. He's a scary one, because he's got a vendetta against me and has more resources than I ever did.

Sure, some of my so-called friends are in high places, but as of NOW they've released me into the ether to do as I will.

Yeah, they give me this crappy work assignment and forget about me, like I'll be just tut with it.

Being a poverty stricken sideshow freak, I know I can never go back UpTown, because some BayBay from my previous life there may recognize me. (Or, one of me!)

When I left Rehabco, I was a gemmat assigned to work at

the Calava Corp. After accessing their training video, I knew I couldn't deal with the utter futility of watching a machine work. I wonder how many PHDs are stuck in this same futile role.

"If I'm ever enthralled by living like this, I'm done for," I thought to myself.

The only possible way to keep your sanity while working for the Big Four, is to entertain your brain with Senselle 'gramming. The only reason a lowly BayBay like me can even afford the Sextel Network is because I have a secret sponsor.

It's supposed to be a big mystery to me, but I think it's either Gramps, because he felt pretty guilt ridden at having been part of the process of rehashing me and putting me back together, or it could be Cami.

With Cami, the motivation isn't guilt. I think that she honestly cares about me. Of course, I know very little about my new beginnings, they didn't tell me much, but again "like a babe in the womb," I absorbed a lot of information they hadn't intended to pass on.

They all carried on their business in front of me thinking that I had no concept of what they were saying – what the intent of the language was. Even though I was not fully formed and wasn't able to, at that time, understand what I was hearing, but I did hear it. Somehow I intuited the feelings, and thereby the intent behind their words.

I think because I was in a jelly-like substance, I absorbed the sound waves, and with them all the information that they had accessed while in my presence.

At any rate, Cami may have given me a lifetime subscription to the Sextel Channel because she feels pity for me. She saw firsthand what the reconstruction did to me.

I don't know who the hell I am, or what I'm going to do with my "new" life, or even when I'm truly going to begin it.

However it was bestowed, the gift of entertainment is a welcome diversion to my dilemma. My favorite three 'grammes are: Esthana Thlehiya's Ritual Circles, Auntie Vee's original programming, "Echo's Corner" and of course, Cami's "Infamous

Road Trips".

I wasn't particularly spiritual "before" my recuperation, but I do like tapping into Esthana Thlehiya's Ritual Circles. She's a Hopitu Shinumu priestess who practices a peaceful, forgiving, gentle religion. Although in the classic sense of the big R word, it's not a religion because it has a pantheon (many gods and goddesses), and rather than humanizing religion, the spirit of nature itself is honored. The spirit of life dwells within us, as with all living entities.

I also regularly access Auntie Vee, who exudes this wonderful earthiness. She talks about the men in her life with humor and acerb wit, she relates old memories of her mother's cooking, and often dwells on the pleasant memories of daily life on Olde Earth.

She's not just all sweetness and light, she also relates some of the rather barbaric experiences that the people used to go through, like "going to the dentist".

I'd like to share part of this wondrous bounty with the other BayBays who live in the Warrens. The problem is, if I flood the local Vidnet for too long a period of time, the viewership begins to overdose on too much rich VR input.

They're used to such small segments, so getting an entire e-pisode is like someone who's starving to death stuffing themselves at a banquet table filled with rich foods.

They wouldn't be able to eat that much, and probably would regurgitate what they did eat, because their systems aren't used to that much food.

People starved for emotional response aren't any different, when they get it all at once, when they ingest pure fear, anger, love, hate or even boredom, it's too much for them to stand.

Some of the PHDs never recover from the emotional overload and rush for the TDM. Their sensual input overload justifies the activation of the machine, and we've lost some good PHDs because of it.

Once I realized this fact, I've only transmitted tiny, miniscule chunklets of free Senselle 'gramming.

I gain access through a generic transmitter (one that's not "credited" to any traceable account), so I'll be able to bounce my signal anonymously out and back through the Sextel Channel.

Apparently, even these little bursts of gramming are creating quite a flashwave.

Her wandering attention was sharply brought back to the meeting. Political commentary over, the communal jack-ins began in earnest. Frankie was glad she had enough chedits left to access an 80% VR jack-in with Arpad.

I'm amazed he's willing to share his technical knowledge of the ROV that he'd recently been called upon to deprogram. It was really funny watching him describe the dangerous ocean retrieval with such earnestness. It was very technical (although I understood it because I had worked for OTEC in my previous life) and there was a lot of it.

This is what it's like being jacked-into him, for our very first time!!

He became animated during his explanation, talking, gesturing, and expounding RE technique. In his own element he lost that ostrich-like awkwardness, manifesting his inner self, a confidently magnificent masculine creature.

In my mind's eye, he is the ultimate space cowboy; tall and rangy, moving with the strength and graceful fluidity of a HORSE.

I can imagine him out in space; scouting out and roping in the wildly careening chunklets of space bound raw materials that are needed so badly on Terrax.

He's like an unsung hero, and just the thought of sharing any part of that journey with him, of him allowing me to see him as he dreams to see himself, makes my heart sigh.

He can't know how much his sharing himself means to me, I'm so very lonely, so isolated.

In some strange intangible way, his energy, his vibe, the comfort level of talking to him about his work, feels very much like being back with Teekum. It's the only thread I have to normalcy.

I mean, one minute I've got a wang and the next minute I'm achin' to have him inside me. It is too freak'n weird!

At those times, I'm too ashamed to let anyone see the real me. The transfer only takes a few hours, but the amount of time between transformations, I haven't been able to peg yet.

I don't know exactly how long I was in Auntie Vee's labs. I don't know when I came out of the recuperation period.

I do know that her voice sounds like an old friend. And, I'm glad for that subscription to the Sextel Channel because otherwise I'd never have the funds to access 'gramming at that level.

Auntie Vee disperses a few of her media releases solely on the Sextel Channel and only the highest paid Nowals can afford them.

Generally, the people who comprise her private audience are the top techies of OTEC (her dear Panas), who are her assistants for the many projects being carried on at her Uberlabs; from developing new technical systems to improvements on genetic engineering. These PHDs, having now been infected with emotions, have changed.

Once the Senselles began running the show around here, the PHDs themselves have changed, because of the Senselles. And the Senselles can't help themselves; it's the only coin of the realm that they have to sell.

Now that I'm having these "feelings" for Arpad, feelings that I never even had for Teekum Dowdy, I know how it feels to be overwhelmed with an unfamiliar emotion. I realize what Auntie Vee was warning us against when she lectures her Senselles on the dangers of "addiction to emotion".

The times I've felt "love" before this was simple harlotry in comparison. Oh, we had sexual passion at (what we thought were) the extremes of human emotion, but that was just the peaks and valleys of the journey, bereft of meaning and fullness.

Now I know the true path to love lies between where you and your loved one connect. You come together where feelings of passion are maybe less intense, but more real. I don't know.

What I felt for Teekum was all the right things – all the things

I was supposed to feel, all the things I was programmed for. I thought I was happy with my life. But once it was interrupted forever, I find I can't ever go back to that.

All I know now are the things that I've learned since then, basically, that at the highest levels of the human condition, there is a shit load of trickery and betrayal. There are traitors inside OTEC, one of the biggest organizations on this planet.

The same human emotions that can spur you on to create the best thing that you could possibly create, can also bring you down to the lowest of lows. Emotions can be your frailty, your weakness. Emotions can be your biggest enemy and your best friend.

Now, I've been infected with these damnable "emotions".

And I know I can never go back to being "just a woman" or "just being a man". Like it or not, I am both.

Somehow, I have to find someone as freakish and outcast as I am who can love me for who I am, for everything that I am. Since Arpad, having gone through the Uberlabs while I was still in recovery, knows who I am and has seen me, and I know who he is and what he wants to be, maybe it's possible we can be truly meshed.

I came to the AGGIES determined to become an AboveGrounder. I'm intent on trying to live in a SadSack, the blobules that grow like barnacles outside of the Hives.
The SadSacks are only temporary, because the harvesting machines come by every so often during in a cycle; they're all programmed for different intervals, depending on the density of Hive incrustation.

The Harvesters are automated, so it's nothing personal. But, if you get caught in one of those when one comes by, you get harvested. You get – dead.

I guess it's kind of an exciting life. I'll need to access enough money to buy spray on Aerosuits, the very best in environmental protection. But, for now, all I care about is that I'm here with all these other PHDs, the wanna bees and pathetic posers, the looky-loos and losers, the one-of-a-kinds and jack-of-all trades.

Here I fit in perfectly among the other freaks. I wonder if we're all freaks of nature. Or, if some of us, like me, were made.

As the meeting progresses, Pavarti, the AGGIE leader instructs us to ignore the Senselle 'gramming that will provide cover noise for our meeting. While Esthana Thlehiya's 'gramming plays for public consumption, the meeting participants are accessing a communal jack-in that will allow our molecules to meld together. As a supreme social experiment, we've all agreed to "energetically" become one during the entire length of the meeting.

Sharing a communal jack-in is against one of the deepest societal taboos, so the excitement on the local network is palpable.

Another first for me this annuation!

I hold my breath, as The Grove 'gramming begins........

* * *

From Esthana Thlehiya's Book of Shadows and Light

It is only eight weeks off from Niman or summer solstice – one of the biggest rituals of the year here at the Rose Parrot Clan.

I don't know why anybody would want to access this mess of mixed up notes on how I organize the ritual. For what it's worth, it's a behind the scenes snapshot of how to prepare for ritual, which some people think takes away from its "magic."

The funny thing is, when you stand with me in circle, there's no pretending, no faking it till you make it. You can feel the power of all the elementals, of the ancient beings, as well as the soul power of all those who stand in the circle with us, magnified by a gigallion while we are spiritually connected as One.

People who don't study physics don't understand that one person's energy plus one person's energy does not equate to only two people's energy!

The multiplied energy is exponential in strength, particularly when you get into groups that are multiples of seven, then the energy is electrifying.

It's more magnified in a larger group than when you stand

alone or are in smaller groups. This is said with the caveat that when the smaller group is composed of particularly intense people, it can be very powerful as well.

Niman is the lighthearted celebration when we bid the Kachinas adieu for the year. It also means we're coming up on the next six months of summer. We turn the Wheel of the Year and feel the physical changes in the world all around us.

As far as the planetary motions, this is the time when the entire of Terrax slowly tilts towards the star that is most radiant and warm in the sky at that time, especially, ones in it's northern hemisphere.

This is the time of year to be jubilant, when we share the work of the harvest and its abundance, as well as the sweet release from winter's toils; the release from study, the release from the inner work that we've just done for the last half of the Wheel.

It's a time when we gather with others in the community; other clans whose beliefs we don't necessarily agree with, but are nevertheless neighbors and friends.

We eat together, talk, catch up on news, and relax. We come together to dance in a spirit of happiness and fun! At this big party, The People will be gathering from all points around the lats and lons of Terrax.

The Hopitu Shinumu Tribe will return to us, each clan member and anyone who has ever circled with us is welcome. Though they may have passed through only once, anyone who came to discuss matters of ancient lore and wisdom is welcome to share our abundance and community. Some of our water brothers came to us for physical or spiritual healing, to mourn the loss of a loved one, or to journey through life's changes together.

We don't necessarily call ourselves a religion, per se. We're not recognized as such, we don't have the same tenets (the belief that if you don't do what we say or if you believe in things in other than what we believe in, that you're wrong and to be despised, corrected or converted.)

We believe in the policy of "Live and Let Live," and that our way is the right way for us – for the ones who claim to be the last remaining Hopitu Shinumu, "the peaceful ones"- but it is not our belief that everyone must think the way we do.

We Hopitu Shinumu are mainly an agricultural tribe. You'd think we wouldn't have that many followers in this technogeek day and age, because I grow crops as part of our daily practice, to remain connected with the living earth.

Our clan attends virtual participation plantings, which are critical to in order to begin to understand life, to understand the completeness of the Wheel of the Year. Unless you can track where you've begun and how far you've come, you cannot see what your efforts have brought you.

Until you have helped a living thing flourish, you cannot cherish those life lessons you have labored to learn. You must acknowledge the medicine or lessons of our animal friends, ones whose souls you've known and loved, whose souls have been bound here in the same space where you worship.

The sacred place in which I worship is called Axtlan, a magical place that the Creatrix Goddess Kokyang Wuuti has promised to The People forever and ever, into eternity.

We stand now, here in the middle of eternity, and we've been allowed access to this beautiful place, that we may experience what it was once like to live on Olde Earth.

Some people think Axtlan was a myth. But, as an ancient philosopher, Joseph Campbell once said, "Mythology, to some people, is actually another person's religion." So if it's not what you believe in, you call it mythology anyway and there will never be enough scientific proof for you that it is valid.

But I was there and I lived it. I can always call upon those remembrances.

I remember my water brother and sisters of the Rose Parrot Clan, so named for their delightful storytelling skills.

In mind's eye, as I look round the circle, I see…Rudmon (the Trickster of the group); She Who Brings Joy to Us All, Brook Dancing Water, who bore the first in a generation of nestlings, Jade Dancing Water (the beloved Goddess grandchild of the Rose Parrot Clan). Here then is the gentle-spoken Rhiannon, whom we have kept in contact with throughout time and circumstances; and there is the lovely, lonesome Rayne, who has come so seldom that we can name it now an official drought.

And then, the inner circle; myself, the leader of the people, Changing Woman, White Shell Woman, Esthana Thlehiya, and my beloved brother, the Bard, Phoenix Gaet. And of course, my longtime love, my mate for life, One Who Sings, who is the strength in my bones.

In all the years we've been together, we have learned the lore of The People, and practiced the rituals of faith year in and year out. We stand fully initiated, we have become the teachers.

That's why we call ourselves, the Rose Parrot Clan. The Parrot is the teaching bird, whose direction is the south, signifying the fire of illumination, intelligence, knowledge, and storytelling. The breath-of-fire is a key element that keeps us all alive, it cleans the blood and keeps the brain mobile.

Our elemental is fire, our primary animal spirit helper, the Rose Parrot.

Today I'm at home, organizing my thoughts for the clan gathering on June 21st, summer solstice. For two months now, the local members of the Rose Parrot Clan have been making masks. We take a spiritual journey to uncover the lessons each Kachina will reveal to us.

Once we come out of journey state, we fashion masks out of local materials. We've discovered these huge, three foot long spiny palm fronds, that fall off the palm tree as they dry up and die. They fall down naturally and we collect them. It takes several years to collect enough masks for the entire clan, because only a few fall every year.

At this time we've collected four masks and will decorate those after we return from the journey world.

We descend, step by knowing step, down to Axtlan to rejoin the spirits of those who've danced with us before. We bring the Rose Parrot Clan members together once more in time and space, from the one-time visitors to elders of the tribe (those who had belonged to original circle of the Amethyst Grove.)

Whether Grove brothers and sisters, or new water brothers and sisters from the Rose Parrot Clan, we invite them all, calling them to us in full voice.

We visualize seeing the dancers returning back here to send the sacred Kachinas on their way with much thanks, reverence and irreverence, happiness and joy.

I'll prepare a dish of food from what we've been growing since before the spring equinox. We've harvested plenty of potatoes and onions, so I can make a nice potato salad and the rest of the food will be prepared by the clan members and guests.

Let's see what is done and what we must yet do before the ritual. We've already fashioned the necessary tools and crafts for the ritual, and made all the individual gifts for the guests. The Rose Parrot Clan will redecorate the Maypod together right before the ritual, (we call it a Maypod because it doesn't have a single beam like a maypole, it has three legs like a tripod!)

The Maypod is our sacred weaving. We weave the fruition of the male and female principals into the Maypod when we bind the legs of our Maypod with ribbon and lace, which we do in the month of May.

As it happens, the rest of the Rose Parrot Clan can't be here with us for May moon, so we'll decorate tomorrow during the daytime. This is only right, considering the month of May is a sun celebration and should take place during the day. In most of nature, the male and female fruition naturally occurs during the day, whether it be cross fertilization, pollenization, or mating. Sexual consummation should be done in the open and without shame, as is the natural way, each to their own.

After One Who Sings and I redecorated the Maypod, we cleaned up the rest of the quarters starting with the beautiful Vigen de Guadalupe. She stands in the east to welcome our guests; She is the opener of the door to all religions and spirituality, the indigenous portal Goddess, She stands in greeting and warm welcome.

One Who Sings worked on the southern quarter, rearranging the stones that mark our Medicine Wheel. He carefully scraped the sand clean of debris, and cut new pine boughs to freshen the air.

I move on to the western quarter, watering the foliage growing all around the foot of our beloved Lady of the Flowing Water. In silence I carefully replace the flowers in her hair, sweeping away the dead leaves at her feet. I pull the weeds growing at the base of the statue, admiring her as she stands perfectly still, head tilted to one side, a live plant twining through the basket filled with flowers. The basket of flowers appears to be resting on her hip, as though newly picked.

Sea shells are strewn all about her toes, along with shiny rocks and crystals, we leave her bright baubles and gems, and braid flowers into her hair. She stands in the West, because she is the nymph of the waters, whose lifemate is the mighty Palolokon, the Hopitu Serpent God of Waters. Palolokon is the God who has to do with all things liquid, from sap to blood to all the waters of the world, he rules the gentle rain and raging rivers.

Palolokon created all the creatures of the water, and wisely chooses as his mate the Goddess of the West, his perfect partner and lover. She stands in her flowing gown, the world gathered in adoration at her feet, as we dance around her in the glorious light of day.

In the North, stands the Lady of Abundance who is Huruing Wuuti. She balances a basket on her head, laden with the fruits of last year's harvest. We're careful to leave some ripe fruit for the birds, squirrels, opossums and other wildlife that come to eat from our beautiful grove. They share in the harvest of avocados, apricots, and peaches as they ripen, each in their own season.

Believe it or not, even this early in the growing season, some plants are already growing in Huruing Wuuti's garden.

Adorned with a fragile necklace of sea pearls, little silver stars and amethysts, She stands in an ample patch of glowingly fragrant geraniums. Beautiful blooms of geraniums, pink and red and one that's striped white and purple. They're my favorite flower because they retain their vibrant color, even as they dry, and smell of cinnamon.

Huruing Wuuti stands in abundance, the tiny tomatoes in the tomato patch behind her are just starting to set, and the potato plants are sturdily growing. The onion sets are coming in, they look healthy. Some of the strawberries are just starting to turn a bit pink, while the others are still green and yet newer ones have just begun to set. They won't propagate for quite a few weeks.

Our garden is growing. We replant every two or three weeks, so that the crops will come up, during the summer, winter, and fall, timing the crops so that we can eat today and also set aside enough for tomorrow. One must always think of tomorrow.

For today, I am prepared for the sacred ritual; I have the lore set into my head, the stories are ready to tell. The dance will be spontaneous. I don't have a dance master as yet, so we will let that happen as it will. We've yet to set up the outside musical components, but we have plenty of instruments for the guests, so it'll come out ok.

Ah, we will see. We will see.

I'm sure everything will come out just as it should. For now, I'm satisfied.

Diary Page- undated

"I want people to know they are not alone,
Instead, when they sit in front of that dark monitor,
dreading to find only a big blank screen,
they will have a message waiting.
They will find me reaching out
to touch their lives,
to fill an empty moment
and stay the depth of sadness."

- Frankie, on Becoming "AV"

Frankie Becomes AV

This is Frankie, checking in. It's extremely confusing for me right now. Things are swirling around in my head, I've got so much on my mind.

I'm trying to look at my situation as though it were a huge story board, using my Vidnet set at 288 mini-screen projections per page.

This way, I see all the different players individually; Auntie Vee and Wad Medani; Gramps and the WC; OTEC and The MOB. Arpad and ...me?

I compartmentalize each player clearly in my mind, each in our own little box. What doesn't make sense to me is that my feelings for Arpad, are changing how I think.

I've never been stirred up like this before. It's strange, because I never had feelings that were emanating from inside of me before I met Arpad.

I always mocked people for wanting "feelings", even though I played at having felt them. I was like an actor, displaying the characteristics, body language, and facial expressions of an emotion, without actually feeling it.

And all of us, whether Nowals or Enbees, UpTowners or DownTowners, whether we see ourselves as BayBays or techies, we're all faking it. Faking it as though our "feelings" are real.

There used to be a line between the made up and the mundane. But once everybody jumped into virtual reality programming, they really thought that they knew what it was like to have feelings. For a moment the PHDs got to be more than crude voyeurs watching through a grimy window. But then, once the Senselles arrived, we were really able to experience other people's memories and feelings.

The Senselle adventures seemed like the best sensations money could buy.

At least, that was before I accessed Auntie Vee's sensory memories. While in recovery, I had access to Auntie Vee's private logs, ones that are only for the TDM, ones that aren't publicly published, even on the Senselles' Sextel Network.

It's not so much that she records the intense situations that other PHDs look for – raunchy sexual experiences or violent encounters- to me, those emotions are like Bollywood compared to real life.

She has more down to earth memories, like the bizarre ritual of "dental visits", which I sneak-peek recorded for my viewing pleasure.

* * *

The Dental Visit

First of all, you have to understand that most adults would put off going to the dentist for years, sometimes decades, these tortuous visits instill a terror that deep.

My stomach turns at the very thought of even calling the dentist.

With slightly shaking hands, I curse my cowardice and dial the number. The oh so friendly, crisply professional receptionist confirms my appointment. Now it is up to me to physically get myself to the building on the correct date and time.

I walk into the clean, orderly waiting room, sign the Sign in Sheet and go to sit in a comfortably padded chair, magazines ready to speed the waiting time. But distractions do not help on a day like today, when first they will x-ray your mouth 12 ways from Sunday, then sit you in another chair to wait some more. The results of the X-rays will be explained to me in detail, but I can't concentrate enough to hear what the doctor is saying, knowing what is coming next.

As they lay me supine in a chair, drape me with a cloth to keep my clothes from getting blood spattered, and shine a high-intensity spotlight in my eyes, they smile and explain that I should relax!

Uh huh.

This proves to be an impossible task considering they crank my petite jaws uncomfortably wide open, and insert into my already aching mandible; two pairs of hands, a terrifying drilling tool, several shiny sharp picks, a suction tube (to clear out the blood and spit) and tiny mirrors so they can watch themselves poking and prodding my sensitive gums.

Then they commence drilling away on my teeth. The sound goes right through my skull and a faint migraine begins around my temples and progresses up to my eyes and down through the back of my neck.

This goes on for several hours at a time, all while the sadistic dentist and his assistant are feeding me questions that are quite impossible to answer. Basically due to the fact that by now I have several pairs of hands in my mouth and will bite off their fingers if I say a word.

After what seems an interminable amount of time, they finish. I feel partially deaf from the sound, and drained from the fear, pain and stress of the last few hours.

They make me promise to come back in six months to repeat the ugly process, knowing full well they will not see me again for at least another five years.

The friendly front office person gives me a bill that is tantamount to a second mortgage, and sends me on my way, smiling with huge, perfectly white teeth.

Once more, I swear off dentists.

Forever!

* * *

Frankie knew she had her audience enthralled, so she continued her gramming. Another rather delightful distraction the Olde Earth humans had were "pets".

To Auntie Vee, her animal friends were as dear to her as any human. Some people who lived back then thought it was wrong to ascribe the same value to animal life as to human life. But Auntie Vee saw no difference at all.

Unfortunately, the animal species did not survive the humans (as they referred to themselves at the time before they "evolved" into PHDs.)

The PHDs procreated here don't believe in the "theory of evolution," we know there wasn't a truly civilized culture before our mechanized society.

And, we simply cannot bring ourselves to believe in this

phantasmagorical world of Nature they tell us existed.

I know, I know, we have all the data; the virtual records of Olde Earth, but it's obviously information that was taken from the explorations of other planets.

We know our planet is too thoroughly desecrated to have ever held life.

And such an extravagant variety of life they spoke of!

There were supposedly millions of species; plants, animals, insects, vertebrates, invertebrates – a veritable banquet of life, offering a massively orgasmic sensory overload.

We are asked to believe that these "people" (whom we are supposedly genetically related to), destroyed all this in order to create what we have here.

That makes absolutely no sense. There's no possible way any intelligent life form would have done that. Ridiculous. An obviously fallacy.

Sigh.

I guess I'm feeling less confused about one thing, at least. My feelings are getting more pronounced. I'm more emphatic about things. Now, while conveying information, it's important to me to make people feel what I feel.

I don't understand this sudden need to be understood.

In my old life, Teekum and I were always running around, frantically trying to figure out what "the ultimate experience" was. We pursued it with all the chedits we earned, and quite a few we hadn't yet earned.

Even with all the chasing, no matter how many bits of sensory experience we could afford to paste together, sadly, we never even got close to the wild variety that was once life on this planet.

I hunger for that variety, and embrace the gramme, especially when Auntie Vee records the minutiae of daily life. In some strange way I think these day-to-day experiences help her Senselles refresh themselves.

I wonder what it must be like, for them. I'd guess if you were from the Other Time, from a place where all this physical experience existed on a daily basis, then to be suddenly bereft of it completely, that it would surely drive you mad.

I think that's why she records personal daily memories for them. I've heard rumors that a few Senselles have begun to go a bit ga-ga, which is completely understandable in this mechanized world. I don't know how they survive the withdrawals from original experience and sensation. That alone could cripple a person's mind.

Occasionally she allows them access to the old rites and rituals she learned from a young Hopitu Shinumu priestess named Esthana Thlehiya.

While Auntie Vee attended sacred circle, she experienced a spiritual mystery, felt the presence of an unfathomable being they believed in, called God/dess. They had million of names for it, and the ancient humans constantly fought over the details, but the fact is that most all humans believed in some form of this being.

It was apparently a supernatural power of some sort, that was either eternally watching over them, or was going to save them, or would grant them three wishes. I dunno, the whole concept of "deity" is bizarre. I simply cannot wrap my grey matter around it.

At any rate, the other thought that has been overwhelming me lately, is that there's nothing left for me to live for except for this deliciously double edged sword that Auntie Vee calls "love."

Even though I didn't have direct access to how she feels about Wad Medani, it's obvious from other reference materials of hers, like when she's talking with Cami (who is one of the few people she knew from their other life.)

She and Cami lived before the Time Rips, and lived to tell the tale. Believe it or not, this is rare, because although the Senselles came here from the same Time, they had arrived from all over the world. Most of them had absolutely nothing in common except this bond of surviving the Rips in Time.

For obvious reasons, the Senselles are strictly loyal to each other, even though some of them don't even like each other, personality wise.

Whatever "personality" means. I mean, we choose our characteristics here. We're all 'grammed and genome-mapped to bits. (Heh, heh – a little computer humor there.)

Here, nothing is left to chance like they used to do it. The old way of procreation seemed like a terrific waste. You would get a lot of sick, broken, violent human beings with that kind of erratic gene mapping. But they didn't think that way on Olde Earth.

You know, our gen-ennges would never allow that. We would never allow such haphazard gene waste to occur.

It is interesting to note that in the OO centuries (the "Oh-Ohs" as we called them) the humans started accepting genetic modification as a safe food process. Major food staples such as corn, maize, soy and rice were already being genetically modified, and (unbeknownst to humans) xenobiology had already begun to be tested.

By the time people realized what was going on, it was too late. Genetic modification was already a commonplace reality within the scientific community, the medical community, and the agribusiness community. The giant GMO food manufacturers; Monsanto, Syngenta, AgriEvo, and DuPont had all banded together to become the first worldwide corporation, The Calava Corp.

That was the beginning of The Big Four Industries.

At this time, humans had several small colonies on their moon. The Calava Corp. chose Ganymede as the site for the first universal wide corporation – the first megalocorp. They now controlled the world's food supplies. This critical industry was owned by Gramps and his old cronies (who would form the Worlds Council prior to the WWWs; Calava Corp would grow up to be the main rival of the Media Overlord Baron's vast holdings.)

Soon, the MOB's textile factories got a whiff of Gramp's way of doing business, and cornered the market on environmentally protective clothing. Thus, the Aerosuit Industries was morphed out of a billion smaller enviro-clothing factories.

I only found out about this because while I was in the final stages of my recovery I was so damned bored, I whacked the lab's history files.

I easily whacked my way through all their feeble security system's defensive programs and quietly snuck over their firewall.

By the way, whacking is how I got thrown into Rehabco in the first place. While interning for the MOB, I dug a little too far into their computer system and they decided to slap my hand by reassigning me to Rehabco for a year.

That was exactly what I wanted them to do, because I knew that's where the Milliner was.

And, I planted a few nanogrammes in the MOB's mainframe on my way out. (They never did realize just what an exceptional grammer I am; even their best technogeeks won't be able to find the source code for my destructive motherboard-devouring grammes).

The 'grammes will hatch long after I'm gone from Terrax, and their "domino effect" will start a system-wide cascade.

Pity I won't be around to enjoy that particular show.

Damn, I'm on fire below!

Arpad. Arpad. Arpad.

Lately, as I try to go about doing my work, as I try to decide what in the hell I'm going to do with myself , with my life, whether I should become an AGGIE or not, all I can do is think about Arpad.

I think about the geekie way that he walks, his bright flashing smile and his almost childlike innocence. Not that he's like a little kid, but he just has an incredible look of wonder on his face when he's talking about space travel. He's absolutely animated when he's telling me what it'll be like to pilot his own space vehicle, describing how it felt to be alone in the vast ocean of universe. Alone with just himself and his wide ranging thoughts.

(Sigh)

It's . . . it's ... just sad. I...I...want to follow him around all the time. I wanna touch him. I wanna be with him. I want to do all these things I never wanted to do with anybody before.

I know it's not too tut – if you know what I mean. I thought that I had cared about Teekum Dowdy, back in the old life, but even as I speak, I'm morphing back from a woman into a man, and I won't be able to see Arpad again for awhile.

Maybe it's for the best, until I can figure out these damnable transformation cycles. If only I could learn to control them!

Arpad's out on a space mission of some sort. He finally got his damn Spav-V deep space vehicle ready, so he can go out and gallivant around in space. Like a half-crazed knight-in-shining armor, a real space cowboy.

Oh... my... goodness. I guess I'm worrying about him.

As I sit here, as my voice deepens and I start to become the other, I wonder how in all the worlds will I be able to keep doing this. How can I stand this constant flipping back and forth between genders? What brings on the change?

What on Terrax could shock me back onto the right path? What could make me stick to one gender or the other? I don't even know if it's physiologically possible again.

Maybe re-exposing myself to "the elements" would give me the shock I need to turn me back into being a gemma full-time.

I think Arpad likes that version of me the best. I'm not sure if he could really get involved in the entire process of loving me both as a man and a woman.

Until I figure that out, I can't let myself get involved in that train of thought. It's especially painful trying to think about my future while I'm sitting here, stuck in this tiny scrap of a W.C. cubby.

I'm not just worried about "falling in love" with Arpad, either. There's something else eating at me.

When we were at the AGGIE MeetUp, I felt a "presence" among us while we were communally jacked-in. It didn't originate from the outside world, our sec perimeters were tight. But it felt like a Big Four corporation rep was there in the mix with us. I don't know how I know, it's just that the energy signature was too tight, too controlled. A light touch, like a ghost among us.

I don't know how They could have broken our code seals from the outside, surely the webnet would have alerted us to an intrusion.

It felt as though it was inside the disteldt shield with us, blending seamlessly into the energy mix.

Whatever it was, when I was One with it, it didn't feel like the enemy. It felt like.....being free for the first time in both my existences.

I want to talk to Arpad about "the presence", see if he felt it too.

Which means I am stuck here, waiting around for Arpad to come back from wherever he's gone, (that is, wherever Auntie Vee and Wad Medani and Gramps have sent him.)

Sometimes I think that waiting around on others is simply not enough like living, for me. I have to find another, bigger, more important reason to live.

In the meantime, with what's left of my chedits, I'm gonna purchase the most expensive Aerosuit that I can afford, and have it painted double-thick onto every inch of my body.

What a laugh on the World. I believe I'll go around without a hologram, and no one will even recognize me. I downloaded a rebel channel to find out how to join the SadSacks, and having an Aerosuit is the first prerequisite.

Right now, the best I can hope is that whatever toxins I expose myself to will shock my system back into someone who I can live with.

All this waiting has me antsy. I'll use a few more WC issued Eedia chedits to access a calming program from the shelf candy available.

It's from Celeste, the cat lady.

* * *

elementals #2- sound

busy birds chirping
twits from the tiny flocks of birds
the raucous croak of crows
whippoorwills on the wing whoop it up

a small dog incessantly barking

ice-cream man's song plays over and over
cars humming lazily by on my street
the ocean murmur of the two freeways nearby

a soft susurration of breeze in the Eucalyptus
tree swaying to and fro
the crazy, ever-changing chatter of the mockers
in the old oak tree

a private plane, seeming to hug the earth roars over my head

the sucking sound of a vacuum cleaner

cleaning my neighbor's house
my cat hobbling by on three legs, lame from a fight
creates an uneven one two three-pause-four,
thumping as he walks
the hisses in pain as his leg buckles underneath his weight

the neighbors' children scream and fight
a mother yells at them to "cut it out"

the TV drones on

the Excited clicks of the hummers as these that about trying to
find the last drop of the spring harvest's nectar

the sounds of life............

March 2070
by Kayla Raconteur

* * *

This is Celeste, the Cat Lady, bringing you an e-pisode of "Grove Memories". I want to share a moment of my life with you, so I'll describe exactly where I'm at and what is going on around me.

I'm reposing in a low to the ground, bright blue beach chair situated in the center of the Grove of the Amethyst Moon.

Facing west I can see the Statue of the Lady of the Flowing Waters, a maiden in a flowing gown, grasping a basket of flowers in her left hand and delicately clutching a bunch of grapes in the other. Twined in and around her topknot are ribbons in shades of lavender and pink. I've carefully woven burgundy red roses into the garland setting atop her head. Shells and glass beads in hues of blue and green adorn her feet, an old wine bottle shaped like a leaping fish nuzzles her ankles.

Listening intently to the friendly cacophony around me, the neighborhood has the usual gamut of background noise; mainly featuring lawn mowers and leaf blowers, with an occasional chainsaw thrown in for good measure. There's always yard work and house remodeling going on in this area, a never ending remodeling job, of sorts!

It's around noon, so there isn't much in the way of street traffic or the shouts of the neighborhood kids playing, because they don't get home from school until around 3:00 p.m. That's when it gets busy on our street; the mustard-orange school bus stopping traffic in the street going both ways, until its precious cargo has alighted safely from the bus, into the arms of their waiting parents..

My beloved furbabies, Lil Bit (a one-year-old male tabby cat), Tootsie Roll Pops (my exotic five year-old female calico cat), and Melanie (our five-year old Stabyhoun pup) are all in the backyard with me.

Lil is hunched at my feet, watching the bugs fly around and plotting his plan of attack on them. Tootsie is hiding, rather effectively I might add, under the dappled shadows of our huge Norfolk Pine tree, keeping an eye out for Melanie.

I close my eyes and listen intently. I am at peace.

* * *

"AV" , Hero of the Peeps

Good morning! This is the SUVA network. The most Superfluous Universal Vacuous Audio here on Terrax, and everywhere, universally.

This is your hostess with the mostest! And, I'm here to tell you about the latest flashwave spreading all over the Hives.... and, who knows, maybe even in the Warrens?

Who knows? Who cares? **You** Should!

Hive wide, the newest sensation that is fully tut with the viewership, is our very own, homegrown hero of the peeps, Anti-Vigilante. We dub this PHD our new hero, because AV has single handedly managed to wipe out the most reviled criminal to terrorize the viewership in hundreds of annuations!

You know who I'm talkin' about, The Milliner.

We don't know if The Milliner was another MOB created menace, all we do know is that this....scamp, this rapscallion, had the viewership in a stone cold panic!

This is the self-same degenerate who was viciously, maliciously attacking innocent viewers-at-large. No one was safe.

And, we were forced to sit, paralyzed fearfully in our cubbies, while the lame duck WC stood by helpless to stop him.

Finally, captured and sent away to Retainage, we all wondered if, after "The Milliner's" reassignment, he would be returned to the general population?

Unthinkable.

Nowhere have we seen such horrific crimes, his victims' faces stripped right off their exoskeleton. This was not a holographic image, folks. He was the real thing.

Then, this demonic creature would arrange the skin of his victim's face atop his head, like a hat, thus earning his moniker, The Milliner.

Where could this cold blooded shill have been housed on Terrax? Who would have felt safe with him on the loose – bar code or no bar code imprinted on his wrist?

No one, that's who.

The viewership had wanted him sent to Phaeron, as far away

from us as possible.

Then, suddenly, The Milliner "disappears" only a few day-cycles before his release from Rehabco! The news coming out of the joint is that The Milliner couldn't be found.

Although Rehabco, Terrax is a maximum security facility, one of the safest places in the universe, somehow this PHD was missing from his cell!

We all assume he was killed.

This is BIG, peeps!!!

Let there be no mistake, there hasn't been a "murder" on this planet (or any planet in the entire World's Council) for thousands of annuations. Not since the mythical ancient ones ruled this planet with war and pollution and greed, has such murder existed.

Oh sure, the Historians would have us believe, that the ancients killed each other over scraps of paper they called "money," or bits of yellow metal, called "gold." Preposterous!

Anywho, we assume the disappearance of The Milliner has a purpose, but, mysteriously, no group or individual has come forward to claim the deed.

Recently released Retainees from Rehabco, Terrax, have been closely interviewed. Once properly questioned, they all claimed "It was murder most foul!"

They say The Milliner was professionally eliminated; that he just "disappeared". No one saw him leave. Although we'd love to talk with one of the Guardians on duty at the time, we still haven't max security clearances from the MOB for the interviews.

They will not talk. The Media Overlord Barons have a chokehold on the entire Rehabco system, universe wide.

But, that's another program for another time.

Just know now that all over this planet, Hivewide, a flashwave has started in support of Anti-Vigilante.

Whether you want to admit it or not, you are a hero of the people.

We feel safer, and we have YOU to thank for it, A.V.!

* * *

Arpad has the Hots for Frankie!

This is Arpad, downloading my daily TDM input.

Today, for the first time in my life, I'll feed this stupid machine with utter confusion. Seems only fair since I'm confused.

Yeah, me.

I have emotions that are at conflict within me. This isn't welcome news, since I'm about to go out roping.

Sure, I'm only taking a couple of small practice runs at random bits and growlers, but even so, the growlers are dangerous iceberg sized chunks of ice and metal, capable of shearing your ship in half, as they are torn off their cometoids.

The bits are trickier to capture, because the flight path of flying debris can change so quickly. Their trajectory isn't easy to figure out, even with the OCTS (Onboard Computer Tracking System.

Space debris is unpredictable, the tiniest fragments (called "silver bullets") being the most lethal. They can hit a ship at thousands of rpms, instantly breaking apart spacecraft into a gigallion different pieces.

Flying through all that smack is sorta like threading a needle with TNT. Once you manage to fly through the field of debris surrounding a cometoid and tie the growler down, the snap impact of the berg you roped in can nail you on your landing.

If you don't have the perfect tension on the lariat, the rebound of that growler can smash your ship like a bug.

No biggie.

I'm not really worried about my performance, I've got a lot of experience with roping in the growlers, the bits I just stay away from.

The question here is regarding the space worthiness of my smaller craft.

Initially, we'll be traveling outbound in the big ship. I call her "The Diego." She's sleek and fast, built for maneuverability. And, she's got the power to tow those bergs half way across the Novae Galaxy and back to base.

The tricky part is my new personal pod, it's been especially retrofitted for my lanky frame. This is what we're testing out today.

The other problem is my state of mind.

I suppose there's no harm in some mental confusion, but I wonder if, even though I'm getting closer to my personal goals, I might be going down a path I hadn't anticipated.

I'm falling ...in....oh man, I can't even say it!

I mean, for pity's sake! Frankie is a girl! And once I actually look at her, I realize, she's gorgeous!

She is …I'm completely discombobulated when I'm around her. I can't think of much to say, which is normal for me, around girls.

It's just that, she has a way of making me feel so comfortable, eventually I start to get to talking to her just like she's one of my bots. This is weird 'cuz I'm not that comfortable around other PHDs.

Makes no difference whether a person is a gemma or gemmat, I'm usually more comfortable around machines than PHDs.

I get nervous. I can't believe anyone wants to be around me, freakish as I am (too tall and gawky, like the Olde Earth bird, the ostrich).

But honest to Vee, when Frankie looks straight into my eyes, it's like being hypnotized. I forget who I am, forget what I look like. It's so easy to talk with her as easily as if she was a bot.

And you know what? She wants to go with me to an AGGIE MeetUp! Me!

Hey, it'll be easy to do considering a couple of my various and sundry employers want me to go to one, anyway. I'm not sure if they want me there "planted" within the organization, or if they're even seriously interested in what's going on with the rebel group. It could be a decoy move, who knows.

It's always been hard to fathom my boss' intentions, but as long as they pay me my astronomical fees, I'm not asking too many questions.

I bet you're wondering exactly who I do work for. I won't even think his name in my head.

Yeah, I know your weakness TDM. I found out from Gramps that the one thing you can't know is what hasn't been thought yet.

You might be able to record everything I say and think, but you cannot intuit; you cannot know what I "might" think.

I could lie to you and you could tell if I lie, but if I omit something, you won't know.

Even as you record this, I know it frustrates you.

A little knowledge is a dangerous thing, no?

HA! Stupid machine.

I can sense a consciousness within the TDM, as I can within any type of machinery. Perhaps that's why I can relate to machines so easily. I trust machinery. I understand its spirit.

There's a difference between soul and spirit. Inanimate objects have spirit and animate objects have soul. Not to engage in a philosophical or a theological discussion here, I am simply telling you this because Frankie has infected my soul.

My confusion comes from the feeling of being completely comfortable around her, as though we are the same person. I know that's not possible, yet it feels real.

There's a mystery about this gemma – a mystery that I don't want to solve, so much as simply "be in the presence of".

I'm not comfortable feeling this way, especially for the very first time. Speaking of firsts, how often in your entire life span, among all the day-cycles that turn into annuations that turn into eons, how many times do you truly have a first? The older you get, the fewer times you experience something for the very first time.

Once your perfectly pre-programmed memory is installed, you know everything about your life, from the moment the "cork is popped" to your last dying breath.

You come into this world "remembering" everything you will need to know. Sure, for the first few annuations you experience quite a few "firsts". Then, as your Wheel of Life turns inexorably on, there are fewer and fewer experiences that you have for the first time.

Right now, I think I'm at midspan, but who knows how many annuations I'll last? We PHDs live for hundreds of annuations, because we've got no moving parts. We're hardwired to last; needing no maintenance, preprogrammed to do what we're supposed to do, so we suffer virtually no stress.

When you're doing exactly what you're supposed to be doing, no matter how bizarre or tedious, whether dangerous or safe, violent or non-violent, there's minimal wear and tear on the internal mechanism. We tend to last a long time.

So, here I am, at midspan, falling in love for the very first time! Here comes the gemma of my dreams, with a gemmat's name no less (I mean, it kinda sounds like gemmat's name).

Don't get me wrong, on her it's totally feminine. I'm entranced by her secrets.

I know she's carrying around some heavy internalized pain. She doesn't talk about her past; her pain seems to come from a secret she cannot reveal. Perhaps a lost love. I don't know.

To tell you the delicious truth, there's so much about her I don't want to know, because it's like getting to the end of a good story too quickly.

Even if there's a "happy ending", you're sad because it's over too soon.

To me, she'll be a never-ending story.

I don't know why the Olde Earth people used to call this "falling in love". If I'm falling, it's not falling down, it's falling up. My heart can barely remain in my chest in her presence, my head feels like it's spiraling skyward away from my body.

Yeah, I'm definitely going to try to be with her at that AGGIE meeting. I want to be there so that in case anything goes down, I can protect her.

* * *

Barrage began his input for the MOB 'puter gramming first thing in his day cycle.

As instructed, I'm on my way to the Aggie meeting. I expect it to be quite interesting, but not for the reasons that other's may think.

As an undercover double agent for the MOB, I'm given specific assignments, but there's also a certain amount of … uh…wiggle room.

How shall I put this? Let's just say that they give me enough rope to hang myself with. There are some vague instructions, along with the requisite gramming nanobots that are injected directly into my energy system.

Nice to be trusted, eh? Don't be so easily shocked. I'm also quietly being encouraged to grow in my own way, but without the "official consent" of my superiors. Twisty.

Recently, I've been practicing a form of p-transformational energy experimentation that has proven quite fruitful. I'll not record these into the TDM, albeit that is highly illegal to do so, they are my secret weapons, for the nonce.

This leads me to wonder – how far do "they" really want me to go? How far will they let me go? Will I be out there someday and find the end of my leash tugging at my ankle?

Well, we'll see, won't we?

For now, I'll appear to be doing as I am told. I'm to personally observe the AGGIE meeting; their banal attempts at rebellion, the second rate technology that surely awaits me and, last but not least, the mini-drama of this pathetic attempt at "blossoming romance", featuring the insignificant Frankie and Arpad, the freakish ice chip wrangler.

Ugh.

Arpad Reitan is a wannabe HORSE.

Frankie's some lowdown, no-count white collar criminal, just out of Rehabco, Terrax. The specs on her say she was hired as a MOB mis-information specialist – an MIS is minor league stuff. Then she got into debt gambling, and was devolved to the Drone Halictidae, Northern Warrens.

She and her meshling, Teekum Dowdy, couldn't pony up the chedits to keep that dive paid for, when he was killed in a horrible UVP attack.

Frankie ended up in Retainage for attempted chedit fraud,

served her time and was reassigned to one of the Big Four. That's where the file ends.

No matter how she got involved with these so-called AboveGrounders, my MOB boss wants me to infiltrate the local AGGIE MeetUp via these two drifters.

Seems like a grunt job, there must be another, less obvious reason Asna wants me there. I'll go over and sit in on that bunch of chicks, see what hatches.

The fact that Frankie and Arpad will end up at the same meeting together is insignificant in itself. They're both small time.

Following them is my cover story, because everyone knows that the rebel group accepts only social outcasts. For the AGGIES, being an "underground group" is kind of an oxymoron considering they believe that PHDs should be living above the ground.

(Which everyone knows is impossible. It simply can't be done on a sustainable basis. Believe me, the truth behind the scenes is that the MOB, OTEC and the Mad MM's are all working together on the problem of finding more living space. If we were able to live above ground and survive, we'd be doing it.)

The meeting begins.

Frankie and Arpad connect up and find a private room to insert their wall jack-ins.

As I study their equipment, I begin to admire the AGGIEs simple, yet effective, communications set up. The AGGIES have an old electro-theater that allows each member to communicate privately with their own VR screen. It's akin to a personal disteldt shield, but the way they have it 'grammed, it can encompass a whole groupmind simultaneously.

You can create a safe mind cluster, so that everyone is able to communicate as a single mixed entity, while keeping their own mental identity separate from the groupmind.

We can choose to remain anonymous individuals, or share our innermost, private thoughts, to be laid naked to each other, in a sense.

All in the privacy of our own cubby.

The set up is elegant, yet effective.

My experience here will be more involved than a simple sharing of thoughts.

I'll use piratelepathy, allowing me to mentally "attach" my electromagnetic signature onto Frankie and Arpad's output. I'll be here anonymously, while enjoying free access to their shared jack-in with the group.

To be able to maintain myself, while attempting to attach to another consciousness is ticklish work. You can unintentionally blow the other person away or you can come dangerously close to losing your own individual energy identity signature.

I know of which I speak.

That's where the practice experimentation conducted on myself will come into play. I'm going to attempt an energy transference, here in a totally public arena, for the very first time.

You see, the unique aspects of my 'gramming genotype allow me to partially disimage and turn into pure electromagnetic energy.

While I am in this state, I run a stronger current than most PHDs, so when I connect into communal jack-in activity, I may surge everyone on the circuits.

When in this electro-energetic state, I can "enter" the main grid as pure energy.

I sing along the connector lines of the earth.

As I allow myself to merge with the others here, they can "sense" me too, but can't identify me, as I'm emanating from a power source inside their own network.

Without a warning of my presence, I enter their minds. Suddenly I feel the first resistance, then it slips away as our lives and minds intimately intertwine.

At this point, We are a new entity. We think with a groupmind. The universe and the laws of physics have been altered, in the blink of a byte.

This is all new to me, yet I can sense that we have a name, an identity.

Perhaps we are becoming like the Dark Ones that the ancient

Earth physicists spoke of, the "Dark Energy" that surrounds all living matter?

Olde Earth physics tells us that what can be known is only 10 percent of all solid matter. The unknown, whom they called dark matter, comprises about 90 percent of the energy/matter that exists. This dark matter is constantly pushing inward toward the center of the multiverse.

This is synergistically impossible. If I had to prove the existence of the Dark Ones to Wad Medani and his MagicoMathematicians, I think I'd have a Hawking's time of it without the hard data.

Thus, the purpose of my experimentation is to meld with the Dark Ones.

To become one with them.

To collect my proof.

I'll give the AGGIES this much credit, these meetings have a suspiciously degradation free energy grid. They've achieved transmission signal cohesion. They've obviously purchased top-of-the-line technology from one of the rogue HORSEs who regularly ride out past Phaeron. (Could it be this Arpad character I'm following?)

As they crank up the amps on the WC 'gramming that will hide the meeting's energy signature, and we all settle in to our jack-in modulation, I discover they are harboring a viral infection within the system.

Unfortunately, when the rogue ship's main frame ported with the transference ship, Arpad's ship caught the viral infection and brought it back to Terrax.

It's a stealth disease from a foreign shore. Hard to diagnose and uncertain to cure, the only way to kill the disease is to fight it from within the planet's motherboard.

This is what I think my fate really is. I'm going to become pure energy and live in the TDM. No part of me will remain a PHD.

Hold that thought.

* * *

I was just peeped with a Very Important Message from The Great Fear, dear ole dad himself.

VEE IT! I'm not in the mood to be his errand boy today. I'm more in my techno- jargon phase. He can wait.

* * *

Back to my world.....During the AGGIEs communal jack-in, I'll play it off like I'm a "zero".

I'll be there lurking and listening, but I won't respond.

I'd much rather it be this way. Then I'll be able to practice melting seamlessly into the meeting's infrastructure without people realizing that I've gotten in. I don't want any of them to them surge out, for Vee's sake.

I used to get off watching people surge out on me. No pun intended, but it is virtually impossible to keep the interruptions from flowing in – even when I'm transmitting to the great and almighty TDM.

(I'm even tired of being irreverent, come to think of it.)

The Boss needs more attention, there's never enough for him.

Although I've put out more than a few minor fires, Asna still wants to tell me what to do with my life. And how to do it. I don't know what his big deal is with Phaeron Rehabco, but I'm beginning to connect the pixels.

I'm thinking that my AGGIE infiltration is in preparation for a bigger plan of His. It's a practice skirmish for the real battle coming up.

As if being "his creation" wasn't enough, I feel like I'm being set up. I'm being framed, made into nothing more than a good little soldier, prepared and programmed perfectly to sacrifice myself for The Greater Good.

Who is He to decide for the entire planet what is Good and what is Evil?

I don't know if the race we've created can be judged in terms of good and evil. That seems too much like oversimplifying life, like measuring something that's almost impossible to quantify.

Nothing is a hundred percent. Good and Evil are conceptually pure. Everything in between total good and total evil – that's what reality is and where it lies. That's the fabric of the intricate web that's been woven for millennia.

Finally, the MeetUp's progressing from the more pubic intro's to the more personal meshings.

I just downloaded to their interface, so that my data inputs and readouts will begin at the exact nano as theirs.

Seamless. I'm "in".

There's a goodly number of PHDs here.

I think it's gonna take a hell of a jolt of power to protect all these people from any outside intrusions. Their disteldt shield should be strong enough to shield me. We're safe from outsiders, and they are safe from me.

I'm fitting right in, flowing through their mixed currents.

Anonymity. Precious privacy.

No one will detect that I'm there – that I am who I am – Barrage Rhyzin, bastard child of the great fear and.......? Who?

I don't know who Asna's meshling was. No one will tell me.

And, that's the rub – isn't it?

Could be anybody. Could be nobody.

Maybe I am my own grandma.

* * *

Pavarti, the "head of the AGGIE movement," was agitated this day-cycle. He was distracted and he didn't like the feeling. It's one of those things where you have to be inside someone's head to know how affected they are by their circumstances.

Outwardly, they must remain calm.

Godzeyes, I hate when this happens to me.

My mind gets twisted up in circles and even though I know it's twisted up in circles, I can't think straight. Even though I start on a logical track, mentally I end up back where I began.

Right now, I have a lot of information to process before the tertiary phases of our MeetUp starts, and I can't afford to lose concentration with so many AGGIEs depending on me.

For one thing, I'm concerned about the upcoming space exploration vote by The World's Council, (which will affect the future of all off-planet space travel.) I'm closely monitoring the deliberations as we speak, having stumbled upon an extremely illicit jack-in to their proceedings.

I don't know if my luck stumbling onto that jack-in happened accidentally-on-purpose (did the MOB or the Senselles arrange it for purposes of their own?), or if it was a lucky stroke of fate for my group.

(Being that I don't really believe in fate, I would have to say that someone had meant me to access that meeting.) So be it, I'll be careful to leave no traces of my energy sig there.

I think the Powers That Be are on a mission to stop Auntie Vee and Wad Medani from procuring independent space mining rights, for obvious reasons. It would put a damper on the MOB's galaxy-wide monopoly on raw materials retrieval.

The Senselles have responded by trying to swing the vote through the mental manipulation of one of the top council leaders, Laird. He's tut with the other World's Council leaders, and the viewership respects him, because he looks good and he makes them feel good, in every sense of the word.

The Senselles sent their best "sexual healer" to brainwash him with a mind-bending Odissi dance, performed by the one and only, Damaria Rotiarte.

Again, I'm trying to fight off the circular mental motion of my mind. The reason I'm not focused on the job at hand, is that I am distracted with my newest obsession.

Oyez, I tend toward obsession rather easily. Once I want it, once I've tasted it, and once I've had it, even in my mind, I have to have more.

Again and again.

I refer to, of course, the vixen, Damaria Rotiarte. As much as I hate to admit, the woman is driving me absolutely crazy. I've only seen her dance once, during that "accidental" viewing I referred to a nano ago.

I wasn't even "supposed" to be there. Attending the dance were Damaria, Laird, and his luscious meshling, the Senselle, Randi.

(And she is randy.)

I'll tell ya…"accidentally" getting access to that healing ceremony was both wonderful and terrible. It was a double edged sword that cut both ways. Now that I've virtually "had" the three of them, that's all I can think about.

It's why my mind doesn't have a resting place.

I'm addicted.

I want them again.

Thus far Pavarti had run the entire MeetUp with his unconscious mind and it went without a flaw or security incident of any kind.

But as the AGGIE MeetUp ended, as he severed the communal jack-ins and destroyed the one-time entry codes, he felt a shadow cross his private thoughts.

Although everything appears to be running as usual, I can tell there is an unfamiliar "presence" here in our energy grid.

But, nothing's showing up on the viral scans, nothing unusual on the data reel. The machinery is working perfectly, nothing wrong with the mechanical energy signatures around the building.

Our Megadeth Disteldt Shield has once again protected our individual identities, and we achieved an open, cohesive interaction.

Still, something isn't as it seems.

The members have all been accounted for, their energy signatures matched to the downloaded data feed. However, I did find a faint, less obvious source of energy coagulating within our "safe" boundary.

Intuitively, I knew it wasn't emanating from any of the PHDs who are outwardly seeking attention, like Frankie, and her partner, Arpad (who looks like he may become her meshling.) They have about the subtlety and sophistication a Drone.

I don't know exactly what he does. I think he works for the

MOB. He's just got that MOB stench about him. I don't even know why, but I hate the MOB even more than I hate the Senselles.

We PHDs don't really have a chance at a "natural" life. We spend our entire life-cycles being sucked into the brain of the TDM. The TDM is turning our race into electronic juice, using us to "get off".

See, that's the other thing.

I usually don't talk this way. (It's been so long since I had any physical desires. Living in a world where all physical desires are instantly satiated, you become jaded in the true sense of the word.)

On Terrax, emotions just bounce off of you. You live without knowing such thing as "a lack of."

You won't ever taste that last breath, you'll just evaporate.

You won't ever feel starved, or feel panicked for the safety of a loved one.

We experience little randomness, and almost no chaos.

The "secret" meeting that I witnessed, was a turning point for me.

I asked myself why the Odissi Temple Dancer, Damaria, had been sent to heal the sexual dysfunction of Randi and Laird? I wonder, can this couple's sex life affect a WC vote?

It seemed like a lot of fuss over such a little problem.

And, more to the point, why would someone from the upper echelon (Laird) stoop to having dealings with the priestess class? (Damaria has been seen going in and out of Auntie Vee's labs.)

Vee is the one to watch, I tell you. You don't have to worry about what's happening with Grandpa and his old gang. They're just tinkerers in the machinery. Oh, they had one good idea, the TDM, but what am I supposed to do? Revere a one trick pony for the rest of my life?

No.

No, the new power behind the throne is the Senselle class. They are the ones pulling the strings.

Vee had sent Damaria to help Randi sway Laird's vote. The

"dance without touch" was designed to fire them into a frenzy of passion, as their physical desire ascended to new levels.

It worked.

They craved each other desperately even before the dance was fully underway.

Amazed at herself, Randi, with all her experience and raw physicality, had never before climbed to these heights with Laird.

Randi supplied all the momentum he needed to explode, crashing against her resistance, while draining away his resolve.

After the dance, their illicit sexual union sealed the deal. He would now willingly agree to anything Randi asked of him. He's tight in their corner now.

They got 'em where he lives too, because evidently the poor bastard has a romantic streak.

He actually loves her, and for some reason, this assures he will do her bidding.

Whatever it was Damaria did to them, well – it was spiritual, physical and mental – it affected the whole package.

As I watched them, as I watched Laird take Randi and Randi submit to him, I realized it was nothing more than everything I wanted with Damaria.

To see lust like that in Damaria's eyes for me, and me alone, for her to desire me like that, to know that I am the only one who can bring her to inchoate ecstasy.

Ah, but see how my desire betrays me.

Look, I'm an addictive personality. I know it, and so does the MOB. Asna Janghaghir, the great fear, says fear is nothing more than going into the unknown.

Is he setting me up? Are the Senselles?

I try to teach PHDs to be free. I want to release the PHDs from their burden, from their smallest selves. The only thing that I want, is for people to have the choice to live without programming.

Personal freedom is the only thing on this planet that you can't have. It's the only thing worth living for. That's why I strive for it. That's why I don't want to submit.

Yet, they're workin' away my resolve by distracting me. Slowly and softly distracting my mind. The technique they chose isn't harsh. It's like a lovely little tongue, flicking me in places that I hadn't even thought to be touched.

I don't know if it's my body or my mind, or both, that Damaria's taken over.

Maybe it was serendipitous. I know that it should have been a private healing, it wasn't something that should've been broadcast.

But the deed is done.

Tonight, off Laird will go to the WC meeting and make his decision.

Randi is acting nonchalant. She's continuing her daily 'gramming, describing her days spent at California Dreamin, a fantasy/sex spa where she worked.

The viewership waits, panting in approval.

* * *

Randi's "Randy Adventures at California Dreamin'- the Health Spa where the Customer always Cums first!"

Randi answered the phone by the fifth ring, halfway hoping that it would be the naughty preppie-boy calling again. Her shift was kinda slow tonight. When a customer finally did walk thru the door, he was usually booked with Dolly Whoa!

Dolly Whoa! was the big boobed, blond bombshell who dressed like a little girl, all ponytails and cleavage. (She attracts the old guys, anyway, so who cares about her skells. The hag can have em'.)

The caller turned out to be Dickie, the naughty preppie-boy.

"Are you smoking a cigarette?" he asked.

"Yes." (She usually was smoking, but even when she wasn't, she was supposed to say "yes".)

"Are you wearing lipstick tonight?" he was beginning to breathe fast.

"Yes."

"What color?"

"Pussywhipped Pink, my pet."

"How far down on the filter are your lip prints?"

"I'm sucking on the fiery tip of the friggin thing honey." He was moaning ever so quietly by now, and she knew he was getting ready to blow his wad.

It was juvenile, but it turned him on for some strange reason, and that was the name of the game, wasn't it. He made an appointment for later that evening.

I knew that our session would turn into the usual arm wrestling match, because Dickie's still convinced he can get me to go down on him, (although he's been a regular here long enough to know that is strictly out of bounds with me.)

I give exceptional hand jobs, and can have him squealing like a pig in a matter of minutes. I can work over that tiny dick of his until he actually fantasizes that I'm wrestling with a big plump wiener in my hands.

Gently massaging his balls one at a time, I get them nice and taut, ready to produce wads of jezz. I use a special massage oil that warms up at my first feathery touch, in a cruel mockery of real heat and passion. It even smells like jezz.

I tug on his dick with the tips of my fingers, drawing on it, stroking it along its length, while cooing to my skell.

I tell him how magnificent he is, my fingers tugging his hair with my other hand, gently bending back his face as if to kiss him, simulating the love play a sex partner might offer.

I have more tricks, many more. I can make his erection last for almost too long, until poor Dickie's beggin me to bring him to explosion with my greedy hands.

I know the secret to any great orgasm is to go slowly, then bring them to a heady, violent rush of cum.

God they love that.

But tonight, my life changed forever when "the terrible twins" came strolling into my limited sex life. After them, I'd never refuse to do anything. To anybody.

They broke my inhibitions, and that's a one way trip, sugar.

Sands and Mandy

When Sands strolled into the spa, I thought "I must be peaking on that tight dope I just smoked."

I thought I was having an erotic daydream.

He was simply the sexiest man-child I had ever seen.

No older than 26, he was 6' 2", with a milk chocolate complexion, and a lean, well-muscled body, that he'd draped in a magnificently tailored camel-brown suit.

His dark, curly brown hair was shoulder length and carefree, framing a face that could have been sculpted by one of the Masters. His large brown eyes were expressive and sweet, with a touch of sadness hidden beneath those outrageously long eyelashes, outlined with bushy eyebrows.

Any woman would die to have those eyes.

He smiled softly and walked directly up me, looking into my eyes with an assured expectation, to which my body, mind and soul immediately responded...YES!

I'm no rookie, I've been in the sex industry for six years, and have seen many a beautiful body come and go. Don't get me wrong, some of the skells were nice, some were even gorgeous and nice, but none of them had managed to interest me personally.

I always managed to keep my personal feelings outside of work.

But, this man was an incubus in human flesh. From the moment I saw him, I knew I would have given my soul to be with him. Unfortunately for me, he had a more hedonistic goal in mind. He had no need of my soul or mind, he simply wanted to enslave my body to his desires. For the night.

(For a lifetime, damn him to unholy hell!)

As I swayed toward him, my breasts almost touching his chest, he turned slightly away, while reaching behind me with his left hand and firmly grabbed my butt. Drawing my body slowly toward his stiffening member, he gestured grandly with his right arm, pointing toward the door.

In the next surreal moment, the most perfect, petite cotton-candy version of Pamela Anderson jiggled into the room and changed the demographics. He smiled warmly down at me, which had my knees weakening.

She snuggled into his other arm in an intimate embrace, and I knew she wasn't going anywhere.

With a tiny catch in my throat, I realized that Sands and I would not be "playing" alone. I was soon to discover that his was an addictive, drug hazed, starved love ritual that would involve breaking down my personal taboos, one-by-one.

It would take a team effort.

Gazing up into those all-knowing eyes, I knew I wasn't going anywhere either.

This wasn't going to be rushed. They had paid the spa ahead of time, arranging three drunken days ending in a spa-wide orgy that would completely release my inhibitions.

As Sands and Mandy lead me by the hand to one of the group rooms, it was as though they worked here and *I* was the skell.

I felt like a child as Sands laid me down on the bed, undressing me one piece of clothing at a time.

Like some newbie, I wanted him to leap on me and start pumping me, but he didn't remove a single article of clothing. Instead, he held me still as Mandy slowly tasted me, her loving lips sensitively working over my tits.

She tongued me like a woman wants to be tongued.

Mandy understood that I wanted her to start at the base of my breasts, working her way up to my nipples (without biting me). I hate teeth gnawing my sensitive nips. As she took my nipples in her lips, she began gently suckling, as if a babe was milking my breasts.

I was getting turned on, and not one bit ashamed that it was at the hands, er, mouth of a woman.

That wouldn't be my first surprise.

I was beginning to get wet between my legs, and I wanted to masturbate, badly. As my hand right hand strained to touch my vulva, Sands restrained my hands, letting Mandy finish the job that she'd started.

The hot touch of her experienced hands continued to traverse slowly down my body, pulling and pushing, teasing my belly, hips and then softly turned to my waiting, warm mound.

I held my breath. I'd thought I only wanted to have a man's mouth on me, but I couldn't believe the rush as Mandy made tiny jerking motions with her lips, brushing them tenderly across my cunt.

I gasped in lusty response as she suddenly started attacking me with her wickedly long tongue.

Skillful and clever, her tongue probed slowly at first, then began to work quickly in and out of my vagina. Practiced and patient, she softened, then hardened her tongue at will, as no man's dick could do.

My first multiple orgasm took me by surprise, coming quick and unannounced.

I cried out in pleasure as I heard a soft pop, first in my head, then from the center of my yoni. It couldn't have been more perfect. Suddenly, as she smiled at me, I fervently wanted her to grow a dick and stuff it into me.

At that moment, she stepped back with a smile that hinted at such possibilities, later. She and I had just begun.

She laid back on the divan and pleasured herself as she watched Sands hold me down with the weight of his naked body and dry hump me from behind, until I came again.

Leisurely turning me on my side, he massaged my buttocks and back, relaxing me, while Mandy started to suckle my tits again, driving me crazy to have her or him or them inside me.

But it wasn't to be. At least, not at my bidding.

Over and over again, they made me cum in ways I'd never imagined that I'd enjoy. And still, Sands wouldn't give me that magnificent dick.

He was teasing me unmercifully, how long would he make

me wait?

As if in answer, Sands motioned for Mandy to undress him while I watched. My hands toyed with my cunt lips, which were on fire for him. It seems this was karmic payback for all the skells I had teased in the same way.

Had I really teased my skells to this extent? My attention was riveted as Mandy began servicing his sturdily wide and fashionably long dick, which swung heavy and full, waiting to pierce me inside. Within minutes, she had him moaning in agony, and even as his eyes rolled into the back of his sockets, he held onto his jezz.

I cheered him on, selfishly hoping he'd save every drop for my now crazed vagina. My pussy was desperately empty, yearning for cock, needing him to fill me up, bursting me to overflowing.

He kept promising me he would, over and over. But when? I couldn't wait another moment! I wanted to force him to fuck me.

Little did I realize that once he started, I'd wish he'd stop. And once he stopped, I'd wish he'd go on forever.

His love turned out to be a blessing and a curse, it was the sweetest torture, the perfect drug. It was heaven and hell.

After Mandy had prepared us both again (the woman has stamina), Sands turned me toward him and began to tease me, his thick cock poking gently into me by the tiniest bit. Then he'd pull out again, stop and push in again another inch, penetrating deeper and deeper with every stroke.

Slowly, oh so slowly. I wanted more.

He hadn't worked his way even half-way in when I came in three sudden orgasms, short waves then long waves, wave after wave until I thought I'd died and gone to heaven.

I wanted to fall into a deep sleep, fully replete with his cum dripping out of me, but he wasn't nearly finished.

He gestured for Mandy to prepare me again, when I escaped from her grip and flipped us over with such violence that all three of us fell off the bed onto a crumpled mess on the floor.

Luckily, he fell directly beneath me, his proud shiny-wet dick sticking jauntily up in the air. Instinctively I leapt on it like an animal on fresh kill.

I landed so hard he cried out in real pain and surprise. Then, apparently growing pleasure, and more pain, as I sank down hard all the way to his root, yelling like an animal.

I was unstoppable and fierce, and I rocked him hard, then harder and harder as he first resisted, then finally submitted to my lust.

At first, he made to push me away, then realized the sexual power in submission, and enjoyed the hard ride I gave him, until I made him cum.

I felt like he was mine!

Taking charge, I signaled Mandy to get him back up again with one of her expert blow jobs, pronto. As she leaned over him taking his cock in her sweet mouth, laid on the floor next to them and I started to suck on her lusciously ponderous tits.

I never knew what I'd been missing before, got a little greedy, and wanted to suck her dry. Those babies were real, not implants, and I enjoyed myself to the point of distracting her.

Mandy hadn't been "distracted" like that in a long time. She was having a hard time breathing, because she was starting to get off while almost choking on Sands swelling erection.

Seeing him start to respond to Mandy, I ripped her lips away from his sticky member and started savoring the delicious mess. If I could bottle that shit I'd be rich, it's the sweetest cream in the galaxy, be it from man, woman or beast.

I slavishly sucked him dry and then, decorated Mandy's cunt with the stuff from my mouth.

Boy, that got her going. As he lay limp, recovery from our double assault, we girls suddenly realized that we didn't need Sands at the moment.

Mandy had brought this contraption with her, she called it her "cowboy spurs". It was really a strap-on dick that felt incredibly life-like, engorging and deflating like the real thing. While Sands recovered himself for another round, we told him to go get us some refreshments from the spa's cafe. As soon as he left, she started working me over again with her lips, smiling a secret promise of ecstasy.

She delivered on that promise, first with her famous tongue, then, ride em' cowgirl!!!

I was almost sorry when Sands came back!

* * *

Jerking himself out of the sex haze frothing through his brain, Pavarti tried to banish the memory of Randi's wild, insatiable lust.

Damaria wouldn't have worked in a dive like that, he thought. Although she did come from a time period on Olde Earth when the temple priestesses were interactive with their worshippers (meaning they physically exchanged bodily fluids), she would never have participated in such an orgy.

To us PHDs, "physical sex" is the greatest evil, because here it means certain death. No one would even consider it.

As meager as our lives may seem to the Senselles, as pale a reflection of what was real at one time, it's the only life we've ever known.

Like any other animal, we embrace life, we desire life. We want to extend life, do whatever we can to make the best of it.

How could we envision living OUTside?

Like fish who've always lived in water, we can't imagine a world where we can breathe air.

Damaria, I fervently hope that I was supposed to witness that "healing" only as a means to help me break up the power of the WC, the MOB and the Senselles. I pray you aren't part of a larger plan to move against me and my AGGIEs.

Even if that turns out to be the case, I cannot turn aside now, it's too late for me. I must tear open their corrupt inner workings and show the general viewership that the WC has been bought. That the MOB and the Senselles are unduly controlling us through their influence on the World's Council members.

I suspect that Laird's not the only one who's been "influenced" by the Senselles.

The Senselles have many effective weapons. They offer the experiences and knowledge of a multitude of people, humans with superior spirituality, intelligence, and culture, as well as humans who are simple commoners (like our own BayBays.)

Still, to me, Olde Earth seems like a mythical place. The Senselles would have us believe you could freely breathe the air OUTside, that clean water fell freely from the sky, and that the Earth itself was exposed to the open sky. They tell us the Earth was capable of producing edible substances, and claim the world was filled with beauty, filled with all manner of living beings surrounding you, all at once.

Incredibly, they'd have us believe that when you came into physical contact with another human, you wouldn't be damaged.

The capacity to smell or touch or hear these wondrous workings of nature seems like a nestling's dream to us.

I mean, we can replicate some of these feelings, but it's a fake veneer, compared to what they claim once existed. Then these hapless humans were supposedly catapulted into our Time, and here they remain.

The Senselles share their past life with us the best that they can, and crave the reality they once knew.

I pity them, and at the same time, I envy them for once having what I can never have. And I hate them unto death for knowing what I can never know.

Knowing I will never touch Damaria Rotiarte, exotic Odissi Dancer of my dreams. Now that I've seen her dance, I understand why mere mortals should not see it.

I'm helplessly enslaved to her now.

And, although I have no right to, I know I must see her dance again.

I must.

* * *

Frankie and Arpad had just come from the AGGIE MeetUp. Frankie said she wasn't feeling too good. She and Arpad decided to go their separate ways, and hook-up later.

Frankie had a special trick she didn't necessarily want to share with Arpad. She was secretly able to use the worldwide connector lines to instantly transport herself to her chosen destination. She didn't need the tubes and shoots that all the other PHDs used.

But that was going to remain her secret.

As she began castenading to her hostel, she reflected on the events she'd just experienced at the AGGIE meeting.

I can't put my finger on it, but something about the flavor of the energy grid there was hinky.

It felt different than any other communal jack-in I've ever experienced. At first I felt fine, everything appeared normal at the start of the meeting. But once the disteldt shield was activated, there was something weird in the energy mix.

It left a metallic taste in my mouth after the jack-in ended, if that's possible. I don't know what it was. It left me feeling slightly queasy and, I dunno, nervous.

I feel like something's up, but I don't know what it is. I didn't completely trust anyone there.

I don't know who this Pavarti guy is, he could have rigged the jack-in to pirate our brainwaves for his private 'gramming needs. Anything's possible.

I need to KB alone and mull it all over.

Ever since "the incident", life hasn't been the same for me. I haven't found a rhythm yet, because I don't know how much time elapses between my cycles of morphing from one gender to another.

I don't know exactly when it's going to come on, except for right before it does. I have a funny pre-change feeling that I'm beginning to be able to identify, then I know the change will be completed within two or three cycles.

Now, here's this "other presence" in my brain. And, as I said, it started during that AGGIE meeting.

For me, the possibility of joining their group, despite the dangers, is strong. Of course I know that living OUTside isn't possible. The only reason I'm considering joining is, where else can I go to fit in?

Look at what I've got to look forward to. I'm a freak. I don't want what society has to offer me.

I can't work in the Big Four factory they stuck me in – the Calava Corp. It reeks in there! There's a stench that I can't identify, a palpable odor that's sickening to have to smell for one nano, much less the rest of my life!

I spoke to Arpad about it and he understands my need to be myself, even at great personal cost. I mean, he's an outsider too, in his own way.

I tried to access him during the meeting but he seemed to be preoccupied. Not that I know him that well, but I think that he was operating on several levels. He was there for more than just info on advanced AGGIE teckware.

I think I love him, whatever that means.

Of course, I remember "love" from my life before, so I do know what it means, but with Arpad it's different, because I'm different.

I'm neither male nor female. I'm both. I don't know if he can accept that. I don't even know if I can accept it. And, I sure don't want to tell anybody, cuz I can't trust anybody.

I know that the HORSEs, although a maverick bunch, work under contracts that are held by the MOB, but he isn't one of the MOB.

No way. No one owns Arpad Reitan.

I think he's in it for the damn adventure, if you can believe that.

A born pilot, he'll figure out a way to captain his own spacecraft someday. He's drawing slowly closer to his goal every cycle, so I know he's gonna make it.

He has big dreams. Actually, they're not dreams; they're plans. He's steppin' right through his plans, one day-cycle at a time.

What stamina! He keeps up with his MOB bosses, and is taking it upon himself to build a special spacecraft. He works triple time; for OTEC, for the MOB and for himself, using each project to further his own goals in a multilayered course of action.

Of course, the raw materials he retrieves will be sold back to the WC. Where else could he sell the H^2O and scrap metal he'll be ropin' in?

For me, I feel like there's somethin out there, somethin even bigger that I'll have to face.

There's somethin coming toward Terrax, it's spiraling inward, coming toward our energy. I can feel the big Wheel of the Year turning deosil.

Finally, it feels like the correct direction for the energy to be flowing. For too long, our energy grid has been spiraling outward, widdershins. It has caused our people to disseminate, to become thinner as a race; becoming taller, thin and fragile.

We've become physically attenuated, due to the weak gravity field showering us from the blue-green star we're circling around. Our sun has given us a grey-green tinged skin due to the wan light of our dying star.

Actually, Terrax isn't precisely circling around the sun; our orbit is more of a wobbly shape. Basically we're wobbling around a great blue-green star, that doesn't give off much heat.

The organic material of the Hives magnify the paltry light/ energy coming into our physical plane from that star, providing us the necessary heat and power.

But, there's something else wobbling around out there. There's something crowding out the light particles. Something that hasn't been quantified yet.

Don't scoff.

Oh, it's out there. Physicists have known about the phantom energy (P-energy) that has been roving around the multiverse, for eons.

It's been labeled sporadic chaos, phantom energy, P-energy, dark energy and dark matter. Call it what you will, it is that which isn't known, that which can't be patterned, assessed or quantified by mere logic, that which cannot be measured by science.

It is the unknown and the unknowable. It is also what creates the motion, the very movement that makes everything go round.

Something tells me there's something about the Dark Ones

I will know more about in the very near future. Still, the "Dark Ones" have nothing to do with my queasiness today. That's a completely separate issue. As though, even among outcasts, there's an enemy among the AGGIEs.

Frankie came back from her ruminations in time to land at her exit portal. She trudged wearily to her hostel cubby and fell into a gray, dreamless sleep.

* * *

Barrage took stock of his mental notes from the AGGIE meeting he'd just attended.

Pavarti, the head of the titch, seemed organized and rational in thought. He allowed his PHDs their chaotic energy, let their individual thoughts crash together. The participants were easily able to access multiple partnerings or interact as individuals.

We were all safely meshing our sig lines behind their tiny but efficient firewall net.

For an underground group, the thought patterns there weren't all that radical. Of course, at these MeetUps, there are always a few of the socially passive; others are angry at governmental control. Some just wanted the high of a mega communal jack-in.

All this social improvement is fine, however, with my personal experience in the OUTer environment, I've come to the conclusion that most PHDs need to live inside and be controlled.

Not control our thoughts, but certainly we should be prevented from breaching the protections against the environmental toxins. At least, until the planet isn't as poisonous.

I keep thinking, the AGGIES can't know what they're up against, they can't know what it's like OUT there or they'd never even pretend it was possible to live AboveGround.

Yet, they talked about the SadSacks with a familiarity obviously based on experience. That's why I know that some of them have lived OUTside in the sack like bags that grow on the OUTside of the Hives. The SadSacks appear to be composed of microbial patches of a viral infection that attaches itself to organic matter.

They described barely escaping the huge machines that harvest the SadSacks, the dreaded Harvesters. The Harvesters were designed to scrape off the sacks, because they inhibit the energy generation for the Hives.

The material is harvested for reuse, because on Terrax nothing is wasted. In a world were there is a dearth of natural resources, any raw materials are grist for the mill. Most raw materials, including potable water and scrap metal, come from space recovery missions.

Ninety percent of the recoverable materials come from off-planet, because we haven't been able to harness any of the materials of this planet for eons. They've all been melted into the goo that became our oceans or buried so deep within the Earth, we don't have the resources to dig them out.

We do have the Hives and the Warrens and our entertainment. All the viewership has been lulled into believing the energy supply will last forever, that it is a constantly renewing source and therefore we have nothing to worry about.

But, I've been in OTEC's sacred Uberlabs and I know, by Vee, that the machinery is starting to poop out on us. It's not about the energy source, apparently.

The problem with machinery is that even with very few moving parts, it will eventually begin to wear out. We (the MOB) have had to go further and further afield to get our raw materials to keep the communications equipment operating, working with materials retrieval operations in environments that aren't quite so toxic.

There're no major plans to coordinate our powers with the Senselles or the MMs to solve the bigger question. No one's talking about that yet. Even the World's Council has no concrete ideas about how to solve our critical lack of water.

I don't really know what hope our little planet has, don't know what, if anything Auntie Vee and Wad Medani have in mind. WHATEVER IT IS, THEY BETTER HURRY!

* * *

On the other side of the planet, the MOB's genetic engineers were preparing to release a new sample of their super distilled Ultra Violent Programs.

Even as Arpad and Frankie attended the meeting and Barrage melded in with the crowd, making them all slightly uncomfortable, but not completely nauseous, we were working our magic.

We've been on a release and record campaign. We release a SD-UVP into the presence of the viewership to see if there's any way to remote control the beasties once they start their bloody rampage.

Only Time will tell if we'll ever be able to get a handle on them. But, we have to start somewhere, eh? It's not an easy task, coming up with the next generation of ultra evil. Especially after The Milliner. He was a classic.

We thought about it and thought about it. What could be done? What could possibly be more terrifying than the havoc our UVPs have already wreaked on the viewership? What could be worse than physical violence?

Once we researched the horror.docs, and witnessed all the horrific things that people have been known to do to each other, we discovered that the only thing worse than actually suffering the torture, was not to see it coming.

No warning at all.

We began to toy with ideas. The more juvenile gen-ennges posed an invisibility gene, while the techno geeks wanted to go with a remote control nanogene that would slowly change the UVPs interior genetic sequencing.

In their model, this would cause a holographic effect around UVP that would work as anti-shielding device. Without actually having a force field around them that can be detected, they'll be internally re-sequencing their gene pattern. In other words, the UVP could see the victim, but the hapless victim can't see them coming.

We got the whole idea from our physics guru, Little Hop. He was the one who thought we could simulate Dark Matter and use it to our advantage. To Hop, dark energy, dark matter, wild energy – that which can't be seen- has always equally fascinated and frightened him.

He thinks that the most frightening moment in an attack, is the second before you actually see what you will face. You're actually more terrified before you see it because your mind starts trying to figure out how to deal with this new reality.

It's the unreality. It's the not–seeing. It's the waiting for the first blow that causes the greatest moment of fear.

That is why the next violence monger we release is going to be the most frightening. The fascinated viewership will be forced to witness the violence happening to some hapless PHD, without the vic seeing where it is coming from!

Nothing on the monitors. Nothing from the cameras, which are literally everywhere. No data will be recorded , so the UVP will get away with it.

It's going to be a bravura show.

And nobody, but nobody, will be able to stop this new menace.

5

Freak of Nature

I'm a freak of nature
I know I don't fit in,
it's been like this forever
my story's....
of freedom
and sin.
My choice of religion
has driven me to rags
It's kept me apart
at odds with the pack,
I'm not tryin to be special
I just am.

By Kayla Raconteur

Arpad Reitan here, going through my checkpoints, while readying today's VR simcaster practice. I've got my favorite "Bits and Growler Program" cued, mind ready to rope in a huge growler.

(That's what we call the bergs that shear off the big planetoids as they traverse galactic space. They got the name from the sound they make while being recovered. They sound like enormous dragons with creaky, little-used voices, growling in anger at having bus-sized chunks of ice being ripped from their hides.)

The bits are the smaller ones, from tiny slivers to chunks no bigger than a moplane.

I keep missing the "toss over" maneuver. I realize I'm off my mark today, as I hit the release button too early yet one more time, and tank the stupid game. The Game Over warning bleeps forlornly.

I decide to give it up for now.

The thing is, I've done this a thousand times before. I've worked with this same program, playing against six other players around the planet, in a small league we call "The HORSEmen of the Apocalypse". Usually, I score in the top five percentile.

My biggest problem this day-cycle is that later on I'm going to another AGGIE meeting with Frankie (our second.) As if that isn't distraction enough, I've been specifically asked to attend by a curious fellow of rather high ranking.

I'm told I'm to feign an interest in joining the group. I think one of my bosses actually wants to use me as bait, to draw a tail and find out who else is interested in the doings of this underground group of subversives. The whole scenario is complicated by the fact that I work for both OTEC and the MOB.

As I sit here and let the program run, watching it mindlessly destroy ship after ship, I don't even care that I'm blowing my score.

The game is set on "random programming", which means that if I don't use it for 18-20 nanos, the space debris' pattern will change. The game is 'grammed as though conventional reality changes that quickly, because roping in space is that unconventional.

You may have space junk flying at you, deadly silver bullets singing by; while being caught in a field of diamond dust (tiny fractals of razor sharp metal scraps that cling together in a magnetized field). No one knows whether the field is composed of pure energy, or light and matter, but if you venture too near the field, it can shred your spacecraft apart.

No matter if you're in a space faring starship or a huge pleasure boat, stealth teck does nothing against it, neither does a disteldt shield.

(Sigh.)

I'm distracted from the game, because I've got a lot on my mind right now.

In addition, I'm preparing my body for a major surgical procedure that will shorten my ungainly height- one nanometer at a time. It seems like a ridiculous thing to do, but cryosurgery is absolutely necessary under these conditions. I've got to be able to fit into my new moplane!

I'm working with the best technicians to retrofit the moplane to my new specs.

I've got the three of 'em working separately on it, from labs all around the planet. This way no one knows exactly what they're building. Each techie is building one piece of my customized, made-to-order moplane that I will dub a "space pod".

That's what I'll be using to go rope in these monsters.

The Bits and Growler Program I was working with before is only a tiny sample of what we do to recover oxygen, hydrogen, scrap metal and other raw materials from space. The PHDs in this field are extremely individual – like me – the mavericks, loners, and freaks of nature. We weren't programmed to come within the parameters of normalcy, weren't lucky enough to win social acceptability.

Not us.

The most recent pennings of Kayla Raconteur accurately reflect my sadness and depression, of late.

Now there's something else being added to my mental mix. Something that hadn't been activated before.

Two quarter turns of the year back, while I was hangin around Auntie Vee's labs waiting on my next set of OTEC orders, I noticed something out of the corner of my eye. It looked like something that was "growing" in a tank.

I wasn't repulsed, in fact I couldn't take my eyes off the thing. *I've no idea what that fascination was all about, but the picture of it stuck in my mind.*

It was almost as though I was meant to see it, but shouldn't have seen it, if that makes any sense.

I don't know what they're doing up there. I probably don't want to know. At this level of power playing, I could get killed for knowin' absolutely nothing.

Which is about all I do know, right now.

The other little mystery that's cropped up, besides all the

work assignments (my MOB bosses, my work at OTEC, and my secret shenanigans with the HORSEs), is that I met this girl named Frankie. I don't know what to make of her, or my reactions to her.

Recently, I had to run by the Calava Corp. to drop off some machine parts they needed. (We recently recovered them from the South Ocean Dome deep-sea operations. For some reason, Auntie Vee thought I needed to deliver them personally.)

The first time I saw Frankie, it was there, while she was on a tour of the factory.

We made eye contact, I think.

Slightly shaken for an instant, I felt as if I had met her somewhere before, but I have no idea where. She gives me the chills, but in a good way, if that's possible.

I've always been uncomfortable around the female gender, specifically; but, around her, I don't feel awkward and ugly and freakish and too tall. I don't feel like I'm frightening her, like I do most people. I'm not as awkward.

With that one look, she gave me the feeling that she was like me – a loner. Not quite a vigilante, but at least someone who is a maverick.

We're not following anyone and we're not leading anyone; we're just out here blazing our own trail. We're not lookin behind us, or to the side, to see if anyone is there with us, we just keep walking our own path.

I think I caught her eye.

I have a gooood feeling she'll actually accompany me to the next AGGIE MeetUp, so now I definitely want to go. I want to see her, to see where all this leads. I'm trying to convince myself it's no more than idle curiosity, but I know better.

For sure I'm not gonna get anything done here today, I'm not able to practice my craft without the proper mental focus.

And I know it's important to practice, to get my skills in peak performance mode. For some reason, I feel like what I'm preparing for is gonna be crucial to Terrax, to our future.

But then, don't people always feel that way about themselves?

* * *

Alone again

No one can touch me
for I am in a place alone
of my own making,
they don't know I'm here,
they can't help me.
I am out of reach.

- BY THE GREAT BARD PHOENIX GAET
(WHILE IN RESIDENCE AT WHITEHALL)

I know what it is to be cursed, Auntie Vee thought, not for the last time.

It is to be alone, to have no one whom you can confide in and trust.

I've no one who is mine and only mine, alone together in our secret world.

Not only is everything I say recorded, even thoughts I may hold "private" are being monitored. I feel nakedly exposed to the world.

But this is the deal I struck with the Devil when we Senselles arrived here on Terrax, and from now on we're all stuck with it.

Yes, I made a deal with the MOB that we'd enter our memories and knowledge into the TDM, as do all the inhabitants of Terrax. If only I knew what it would cost us, I may not have agreed.

I didn't have much choice at the time.

I suppose you could say that right now I'm in my blue period, one in which all the songs that I compose under my pseudonym "Kyla Raconteur" are depressingly sad. They express a pathos that reaches into the bones of my being.

I am attenuated, my personal energy field stretched dangerously thin. Wad Medani is far away. He went to the Southern Ocean Dome to check out a suspicious ROV accident.

Before he left, he asked me to "read" the energy grid around the accident. I intuited that part of the mystery would point to Asna and his louts, but warned Wad that it didn't feel like the energy was originating from Asna. He immediately preferred to believe that his arch enemy has something to do with it, and intuition be damned.

Funny, he will act on certain visions I've had, but only if it fit with his agenda in the first place.

Humph.

Despite my warning, he'll go on his own instincts, and for that I can't fault him.

If you're in the field, you call the shots.

Whatever the outcome of their face to face meeting, I can feel a massive energy stirring in the ocean depths of my soul. It is responding to a vibration, an energy field that seems to be drawing nearer to me at a great pace.

In my mind's eye, it appears to be a female energy, shaped like a great sea serpent. Her iridescent scales are made entirely of beautiful ocean shells, gems from the depths, jewels and treasure that haven't existed for millennium.

The shells bring to mind the huge, iridescent pink conch shells that I once had on my western altar at home. I remember, when I'd put the shell to my ear, within it I could hear the distant, soft roar of ocean waves crashing on a beach.

It was the very same beach that the Grove of the Amethyst Moon would walk on, our bare feet sliding through the sand along a deserted cove's shore. Full moonlit nights, we'd journey walk together, hearing the slight crunch of the fine white sand under our bare feet, or the tinkling bells Brooke wore around her ankles, while the trees around us were swaying in the gentle wind. We would approach the cave mouth together, eager to see what lay ahead.

This magical cave would be the beginning of many of our

journeys. You might say that the entry into the cave was an entrance into our own subconscious or unconscious mind, but after being there so many times, I realize it's a real place, separate and apart from mundane reality. It is only accessible through that door of the deeply journeying trance-like mind.

Sometimes when you journey you arrive in another world, perhaps one where you're immersed in water. You may be a human, an animal or another creature; you may be just consciousness.

As you swim through the mercurial water, you'll meet other life forms.

Occasionally they'll be mythical creatures, like the ones you'd been taught about as a child. Perhaps the Mer people – mermaids and mermen; or elves; or the fae.

Other times, it will be entirely new beings that you've never heard of before – neither fish nor fowl. The nameless entities may communicate with you, even though they don't "speak".

When communicating with them, you may not necessarily "hear" words, but in your mind you know what they are saying to you.

Most times, you can project your thoughts back to them. Maybe in some way they understand what you are saying, or perhaps not. Either way it's a new, and undefined, world. One you must get to know through your own experience, feeling more and more at home each time you return.

Suddenly, this part of your journey changes abruptly, as a door appears before you. The door is inviting you to enter, shining softly, a luminous emerald green. As you approach, you see it is cut from one solid piece of emerald!

Triangular in shape, it opens from the middle, beckoning the consciousness inward.

The "deeper" into the journey I go, the less I'm aware of my corporeal body. I can barely feel my body around me.

My hand slides along an alabaster railing as I descend the steps. At the bottom of ten evenly spaced steps, I enter a chamber.

There's no water around me now, as I stand bathed in a golden green light. I feel completely comfortable here, as though I belong here. As though I had been here before.

I simply stand, at one with the light, at the same time surrounded and infused with this light.

No thoughts disturb me, no worldly problems call upon me.

No requirements that I return feelings or empathy.

Simply being, in its most basic form.

I wish I could stay here forever, in the pureness of being.

But, it is not to be.

Slowly, overpowering my will to stay, my consciousness is being pulled inexorably back to Terrax.

To the NOW.

* * *

I see that emerald chamber receding, becoming smaller and smaller.

The emerald door shuts, as if a movie is being rewound awkwardly in reverse. (The image looks like this because I'm being brought back against my will.)

I want to stay in the vision, because the world I have here is devoid of feeling and texture; it offers nothing of the things I used to love. Nothing of the beauty, of the emotion of life, but rather the dull dryness of living all inside your head.

Imagination, vast though it may be, may spin and weave wonderful stories. These stories may act as medicine or lessons to the listener.

Still, sometimes even the storyteller simply needs to sit, and listen.

I wish that I could just sit and listen....... to the stars.

Instead, my Senselles call me back to them. I'm to be alerted whenever Cami is 'gramming. I'm able to transmit and receive covert communiqués to the Sisterhood while she is 'gramming without the TDM monitoring our messages.

Busy girl! She's about to begin a new adventure, so I'll sign off for now and begin IMing Wad.

* * *

The Whys and the Wherefores of My Youthful Defiance

My entire, official, real name is Catherine Alice May (my maiden name) Inesco (my married name.) The name I chose to live by here in the future, is Cami.

I chose the name, not only because they are my initials, but also cuz I change like a chameleon. Get it- Cami, chameleon.

The range of changes I go through would surprise you, looking back on who I was when I left Olde Earth. I went from living like a common house frau, to acting out the wildcat I am inside, because here on Terrax, I can live as BIG as I'm feeling. The more the merrier, bring it ON!

Trying to "act big", boy, that phrase sure brings back the Olde Earth memories.

When I was a kid, my father used to beat on me regularly. He'd yell me to "stop trying to act big."

He'd tell me I was nobody, that I wasn't worth the powder it would take to blow me up.

And when I wouldn't agree with him, he'd wail on me some more.

So, I'd split. I'd run away.

I went out in the world to have my adventures and, because it was his pleasure to continue to beat on me, he would have my ass dragged home by the police.

Oh, yeah, he'd get his jollies off, having the cops drag me back home against my will. Then he'd beat the crap out of me. Again.

Finally, at age 16, I managed to break away. And I never went back.

I went out into the world seeking adventure once I became a teenager, hitchhiking around, checking out what was going on.

Trying to have me some fun.

I've always been fearless that way, unafraid of what I'd run into out there.

I wouldn't wonder, "Am I going to get hurt?"

You know, I didn't pretend to know what would happen.

Maybe some fantastic, magical person would come along and we'd get together, get high, get laid, who knows?

I knew there was a possibility that the scene could turn violent, but that thought didn't stop me.

I mean, I came from violence, from getting hurt, so that held no particular terror for me. Besides, that's a minimal part of it. There's also the possibility of wondrous things happening out there.

I knew what I had at home, but I didn't know what there was waiting for me out in the world. I figured at least I'd have some great stories to look back on and remember when I was old.

But that was only a small part of my old life. My youthful adventures are recorded elsewhere, however, the following 'gramme is an adventure I had in "midlife."

Humans called it "midlife", because back on Olde Earth humans only lived a puny 80 or 90 years. By the time you were 45, you were already at the middle of your life!

Of course, an individual's life span could vary, depending on their genetic make up, whether they'd taken care of themselves, and how well they had fended off all the environmental toxins.

Even back then, deadly environmental damage caused by the overuse of toxic chemicals had been insidiously creeping into everything humans touched, from the food that we ate, what we poured into the water supply, and let seep into the Earth.

Humans didn't understand that the Earth itself was one big recycling plant, called an ecosystem. In our pride and selfishness, we overtaxed its ability to rejuvenate and restore itself. The moguls of industrialized society refused to take heed to the environmentalist's warnings, and, the Earth became poisoned past recovery.

There I was at age 45, (used to living a life filled with exuberance and energy), when my health began to seriously deteriorate due to local environmental poisoning. Some days I was so exhausted I couldn't rise from my bed, even after a full night's sleep. When the condition worsened, causing extreme fatigue, joint and muscle pain, depression and narcolepsy, I decided to go to my doctor.

I was diagnosed with a terminal illness called Lupus. Lupus stripped me of all my energy, of my ability – as ironic as it is- to go places.

I mean, that has always been my big deal. To be free, to be able to jump into my car and go wherever I wanted, whenever I wanted to go.

I wasn't ready to retire from my career. For the last 15 years my husband and I had built a successful solar electric systems installation business. We were working so much that we spent all our time at the company.

Even though we were young enough to still enjoy life, and had tons of energy for work, we never went on vacations; we never did anything with friends. We just worked.

Once I was diagnosed with Lupus, I couldn't keep up the pace. I had to quit my job, minimize the socializing with my friends, and stop enjoying my hobbies. I had to give up all the activities that made me happy.

I used to lead a spiritual circle of like minded people who would gather in spiritual harmony, in an ancient Earth based system honoring all life. (It wasn't based on monotheism.)

I used to organize a small business meet-up group that encouraged new business owners to build their businesses through networking with other small businesses. Unfortunately, all of it had to come to a crashing halt. There came a point when I became too ill to be able to handle the stress of running my company; the household duties of daily cooking, cleaning, and taking care of my family; plus all my other social activities.

I was going into a long downhill slide, becoming more and more ill, when I finally discovered that I could regain some of my health through organic nutrition and weekly acupuncture treatments.

But, all that didn't help my mind cope with the fact that my body had become my physical prison. The thing is, just struggling to feel "somewhat" healthy wasn't enough to satisfy my adventurous nature.

It's true. Even at age 45, I still have that wild streak in me.

Inside I'm still the same teenager who used to run away, unafraid to jump into a car with a strange man, not knowing if it would be my last ride or an unbelievable adventure.

True, I grew out of that phase of my life. In other words, at 45, I was looking for a different kind of excitement.

Not the excitement of new sexual relationships, or the thrill of successfully promoting a band, or creating another business.

Cami and The Street Race

At this point in my life, the only thing that is really chancy and wild is street racing. It's illegal, of course, because it's dangerous, but that doesn't bother me.

It's not like I race at night along deserted back streets, like you see on T.V.

I race on a long, residential street, in a quiet section of town. Sure, over a million people live here, but by today's standards, it's not that crowded.

It's a busy enough street. The streetlights are timed so that if you go the legal speed limit (30 miles per hour), you can usually clip along at a good rate and make all the green lights.

Once in awhile, you may need to stop for a red light. But, usually you can cruise for a good mile or two at a stretch. And, it's usually not frequented by the local cops.

Every so often, when I'm sitting at a particular stoplight on that special street, and if it's not too busy, I'll look around for someone to race.

Especially when I have a good health day, those days when I feel energized, like my old self again!

When I race, it isn't that I have a death wish or anything like that; the simple truth is that I love to court speed in a car.

I love going fast. I've never had any professional training, but I've raced on this street a couple of times, and I know it pretty well.

(I guess you wouldn't understand this urge, unless you're a person like me- theatrical, dramatic, have a flair for danger).

On this particular day, it's around 2 in the afternoon, and there isn't all that much traffic on the street. Not many people coming from or going to work at this time of day.

I get to the place I usually use as a good starting point for a race. This is the intersection is where I "throw down the challenge".

As another car pulls up beside me at the stop light, I glance over at the driver, casual-like. If he's alone and driving a sporty ride, I'll rev my engine a little, smile like a cat with a lizard in its mouth, and nod toward that long, inviting open stretch of road in front of us.

It's obvious to the other driver that there's no blood in my eye; it's a friendly contest, nothing more. There haven't been many times that the guy will take me up on my challenge, it's only happened two or three times.

Thus far, women have never raced me; and a guy definitely won't race me, if his woman's in the car with him. I guess they're too chicken of getting into trouble or a fight.

I can understand the thinking there, I'm always alone when I race.

Anyway, it's not about putting someone else's life in danger. It's for the pure thrill of it!

It's the rush of going as fast as you can – challenging yourself and challenging your car to… win…to get there first.

Most of the time we're both going fairly fast, but there's never animosity, it's all in fun.

Little did I know, today was going to be different.

The first indication as I pulled up to his car (before I even gave him the look over), I could feel the hatred emanating out of him!

I felt it radiating out of him – through his car, through my closed window- in wave after wave of unremitting fury.

I figure, the light is about to change, but we haven't revved our engines, or thrown down the jovial challenge.

I have just enough time to think "he's ready to kill somebody."

I sense an element of danger here that's never been here before.

In a split second, the light turns green and the race is on!

At the beginning of the "run", the street has a tiny dip on the right side of his lane, so I was able to get off the line quicker than him.

Determined, his jaw set like steel, he caught up with me pretty fast. I looked over to see what I was up against, and found he had an older model blue Mustang. It wasn't new, but it sounded like it had a heart of gold.

For his part, he had to contend with my drop-dead, gorgeous, red Grand Am, with only six cylinders, but a hell of a lot of guts.

Off we went, charging down that residential street like the hounds of hell were on our heels.

We enter the windy, residential portion of the street. There are cars parked on either side of us. With oncoming traffic in the other lane, we have very little room to maneuver.

Although it's a four lane street, the cars parked on each side leave absolutely no room for error.

Unlike the other racers I've been up against, this guy was not being careful. He was driftin' over into my lane a couple of inches, losing control ever so slightly.

I wasn't even looking at my speedometer, worried about getting side swiped by the oncoming cars in the other lanes.

Man, we better get to the straight-away, and soon.

I'm not good with distances, but the straight-away is at least a mile long. It's the best part of the race, because you can really pick up speed there. We were hittin' the lights perfect – going faster and faster.

After I jumped off at the starting point, I had pulled ahead of him, but he was pushing his car hard now. He was a little ahead of me. Throwing all caution to the wind, I raced ahead of him, pushing my Grand Am.

All at once, I pulled a fraction ahead of him. Then I leaned on the gas.

Finally, I started pulling away decisively.

That's when you know you're going to win the race, when your car is humming slightly off the road, the wind swooshing by and you don't have a care in the world.

It's the sweet spot. Your car elevates off the road just a tad, and man, you know you've got the guts! Now you're driving just for yourself, challenging yourself to go faster than you've ever dared before.

The thing of it is, at that point, I wasn't paying attention to him, or my speedometer, or even the road. I'm on a speed high.

During my other races, maybe we weren't really going that fast, because I had a chance to look over at the other driver, and smile.

You know, kind of enjoy it with him – because whether I was winning or he was winning, we were racing, man. It's a charge.

But in this case, it was to the death, man!

I mean, we were goin' neck-n-neck, gettin' a little bit out of control, floatin' over into the other lane, with oncoming traffic, and parked cars on either side of us. It was reckless.

By that time, his car was coming into my peripheral vision, somewhere behind and to the right of me. All of a sudden, I see this blue streak out of the corner of my eye, and then he disappears.

I dared a look to the right and found.........he was gone.

He'd peeled off! He was out of sight!

I had WON against the death commando in the blue Mustang!!!!

I was waving my fists in the air in celebration. WoooHoooieeeeeeee!

I was totally high. I mean, I wasn't upset. I wasn't angry. I wasn't trying to kill myself. You know, I wasn't trying to put my life in danger, but man, it's an adrenaline rush that just cannot be beat!

I can't remember for the life of me the last time I felt this... exuberant!

There I was blazin down the street, "wooo-hooing" to myself and having a great time.

Suddenly I notice the next big intersection coming up. There is a humongous accident splashed in the middle of the road. It looks bad.

I can see at least three cop cars and two fire engines there, so I'm trying to slow down kinda easy. Trying not to skid into the intersection, what with all the cop cars, their lights flashing, and sirens wailing. It's a freakin light show we got goin on here.

There's a huge fire truck parked in the middle of the mess, and three cars crumpled up in the intersection lookin like tin cans that've been smashed by a trash compactor.

It looks scary. The ambulances had already taken the passengers away; there was glass all over the place, wreckage strewn in the road...... and one mother-fucking angry cop, staring straight at me.

I didn't quite screech to a stop, didn't lay any rubber stopping, but you could see the cop had blood in his eyes.

The straight-away that I had just raced down was a clear, straight shot, at least a mile long.

I knew I was caught because I had righteously trumped the little blue Mustang, and you could see that stretch of road perfectly clear from where the officer was standing.

When the light turned green, he pointed directly at me with the finger of death. Then he pointed at his feet, expecting me to pull over, like a dog come to "heel".

As I pull over, the last thought in my head was, "Holy Crap! I'm probably goin' to jail."

Then, the miracle occurred. My mind mercifully blanked. For the first time in my life, I was completely devoid of emotion.

I wasn't scared. I wasn't nervous. Nothing.

I knew I had just gotten caught street racing. I had no idea what was coming, but I knew it couldn't be good.

Yet, I remained blank…becalmed…an empty vessel.

"Do you know how fast you were going?" the officer barked at me.

"No sir, I don't."

"You must have been comin' down that street a hundred miles an hour! I saw you! You passed four cars. Didn't you see those other people?"

"No, I really didn't see them."

"Drivers license and registration," he bit out, not bothering to hide his disgust.

I showed him the documents.

"Why in the Sam hell were you goin so fast?" he watched me closely.

"Well, I just came from the doctor's office, and, and he had a lot of bad news to tell me. There's a lot of stuff that's wrong with me..." I trailed off lamely.

"Yeah, like your driving" he spat back at me.

I had a wrist brace on and I stared at it. He just looked hard at me, unimpressed.

"Is your speedometer broken?"

"No, not that I know of, Sir."

"Then why are you in such a hurry?"

"I really wasn't in a hurry. I was just going home and didn't realize how fast I was driving."

"Well, you could have killed somebody! Look at this!" He gestured at the mess in the middle of the intersection. "We just sent two people to the hospital, and one went out in a body bag. You could've killed somebody, or got yourself killed!"

He took one look at the traffic congestion, figured his time could be better spent on traffic control and snorted his contempt at me.

"I don't have time for this today. Just get outta my sight!"

He wove me on.

Off I went, slowly driving down the road.

I was numb for about 20 minutes after that.

I mean, I didn't feel anything – no relief, no fear.

I should've been celebrating my triumph over the cop, but I just drove down the street as though it were a normal day.

As though that hadn't just happened for the first time in my life.

(I mean, I've won a street race before, but I've never gotten out of a ticket before. Ever.)

Especially when the cop just caught me goin' a hundred miles an hour!!!

***

So as you can see, even though I'm 45, even though my hair is silvery, even though there are deeply scored lines on my forehead from thoughtful inquiry, even though I have frown lines, and fine lines around my eyes and mouth from laughing and smiling, my spirit hasn't weakened or mellowed.

I'm still the wild child of street adventures, the daring darling of road races, the maestro of my own life's symphony.

I'm Cami Chameleon, forever.

Actor Without Make Up

The actor sits upon a stage alone,
invited to the party,
but didn't want to go.

We all know
what's been bothering him,
it's too hard to say goodbye.
We all know
that it's going to end up
this same way...

Three months of living with strangers
change them to family,
we are the ones
who dare to be another,
to show both the foul and the fair,
the audience to see.

We are a cast,

known as the players,
but when
the last curtain calls,
we all hold hands
and bow.

The critics scoff or praise, mostly scoff,
some of us go to cast parties,
some take the makeup off.

A playwright endlessly discusses
his favorite rendition,
while the lonely actor on the empty stage
sits
awaiting
his next audition.

- BY THE BARD, PHOENIX GAET

I am Pavarti. They call me the maestro of madness, simply because I advocate complete freedom for my people, the PHDs.

Although we are born cyborgs (part machine/part electrical energy), and we remain only a thin scrap of what used to be considered human, even though we appear to be no more than random light projections (images on a holographic screen), we deserve to be free.

I advocate complete self-freedom for any PHD who wishes to join the Above Ground Movement, or AGGIEs, as we've been called.

In order to facilitate the survival training for the AGGIES, I'm engaging in the first stages of a ritual self-cleansing. In this way, I connect with the hero within myself, and thereby, can lead others to do so.

Each of us will become the head of our own individual AGGIE cell, known as a titch, living within each of the four Major Hives and Drone Warrens.

We have placed cell leaders within the Melissae Hive (the home of the Senselles), the Mega Chilidae Hive (MOB headquarters), the Odissi Hive (the entertainers, artists, musicians and dancers), and the Colletidae Hive (the top editors and content producers on the Vidnet).

We've invaded the Hives; will soon infest the Warrens. We've got two groups living in the Drone Andrenidae (the mining Warrens beneath the Melissae Hive), Melissae being the most critical Hive on this little planetoid we like to call home.

Our goal is to overturn the existing power base; our method, to educate the viewership on how the triad of power; (Wad Medani's MMs, the MOB, and Auntie Vee's Senselles) have been lying to them all along, just to keep them a captive audience.

Some of our people have gone on to live the brave new truth. After being initiated during their majestic firewalk, they can live free, aboveground, without government infringements on their thoughts and actions.

If they survive "walking the storm" they are dubbed SadSacks, and may choose to live as perfect warriors OUTside, for as long as they can survive the elements.

It's true.

It can be done, despite what the Powers That Be would have you believe.

Of course, I'm only telling half truths to the legions of people that will become our "true believers". If they knew the whole, terrible truth, few would even attempt the warrior's path.

How many would then go on to become perfect warriors? Some secrets must be kept.

AGGIE SadSacks are thoroughly trained, then severely tested, before they attempt the firewalk.

First they must be grounded on the severity of the OUTside environment; they must show an intimate knowledge of the physiology occurring on the surface of the planet.

They must prove to me that they are worthy of living, for however short a time, in complete freedom. I have witnessed the needless destruction of many walkers, who have gone OUT unprepared.

When an incoming scatterstorm is announced by the Cheops Weather Bureau (the universal weather frequency), we have approximately ten to fifteen minutes before it will touch down.

We breach the nearest exit portal to the upper world, the forbidden Above Ground, using the brief communications failure it causes to organize the recording equipment.

I usually prepare the walkers for several annuations before they try their firewalk. Only one attempt is necessary.

No one is guaranteed they will survive.

However, once you've managed to survive the most violent storm this planet has to offer, you fear no man-made evil. You relish your own strength and power. You finally understand that the ability to live in the manner you choose is your absolute birthright.

As for the scatterstorms, they are both terrible and beautiful, their long thin ribbons of fire coalescing with the ambient phantom energy surrounding them. They're charged laser beams, randomly caroming around in space, able to pierce any man-made material, including a spaceship. Walking the storm is serious business.

Apparently, the Hives are the only organic living materials that are able to repel the scatterstorms. In fact, their energy signatures spike at extremely high peak rates during the storms.

Luckily for the BayBays, there's no harm from the storms to the Hives growing above ground. The Ocean Domes can also withstand them, because they are composed of the same bioelectric materials as the Hives.

After your firewalk, you can choose to reside Above Ground in the SadSacks, hence our nickname. By living this nomadic life, you have basically offered yourself up as a test subject for the rest of the AGGIES.

We monitor your body functions, this helps us develop new materials that will fend off various enviro-toxins. We learn how the body deteriorates, as the AGGIE constantly moves from Sack to Sack, until we succumb to the Radzeeze, or until the harvesting machinery gets us.

The SadSack is an organic parasite, growing in globular shapes on the outside of the Hives. Unfortunately, they disturb the electron flow of the incoming star matter, and in order that the Hive's power grid charges correctly, they need to be cleaned off periodically.

The Harvesters scour the outside of the Hives on a rotating schedule that we've tracked, so we can forecast where they'll be next with about 70 percent precision.

Generally, the Sacks are habitable for at least a quarter of a Wheel of a Year, so our people have developed social customs and adaptations based on this new abbreviated calendar.

We're not limited to only polyamorous meshings, as all the other PHD's of Terrax are required to have. We can take individual partners or, even more distressing to the WC, no partner at all.

Obviously, being nomads, we own no material goods.

The lifestyle itself is exciting enough to make up for the lack of comfort, believe me! Between trying to acquire breathable air, potable water, and semi-decent prot shots, it's like dancing a jig on a high wire, above a stage without a net.

We've managed to infiltrate the Calava Corp., so now we have unlimited access to "toss-aways" (expired prot-shots), containing only slightly more toxins than WC Allotment Levels permit.

We've even discovered an advantage to exposing ourselves to the light levels of exposure to the enviro-poisons. We found that as we slowly absorb the toxins, we've become immunized to the lethal "toss-aways" of the Calava Corp.

Even with the slight immunity, we find that Radzeeze is exacerbated by OUTside exposure, so the ending date of our personal time-line remains about the same.

We survive.

For as long as we can.

Actually, we've only been tracking our progress with the toxins for a few Wheels of the Year. Unfortunately, most of our meeting places have been discovered, so it's hard to retain meaningful records.

Traditionally we had conducted our titch meetings in the

dilapidated Warrens; primarily, because since it's extremely hazardous down there, the BayBays won't follow us there.

The MOB's spies are painfully obvious in their attempts to infiltrate the group because of their unfamiliarity with our customs.

Of course, there are those few instances when the relatives of the "walkers who didn't make it", have turned us in. Their meshlings blamed the AGGIES for their partner's death, rather than face the painful fact that the walker was not a hero – was not a perfect warrior.

Without the teachings, without the self-discipline, without the willingness of self sacrifice, one cannot walk the fire and make it through whole.

I realize, as I delve into my inner being, there's a lot to this.

First, I must calm all the machinery in the building in order to still the noise within. I always begin my self cleansing ritual with my exterior physical environment.

My living space at present consists of several empty cubbies, which are empty due to the severe structural damage to the Southern Ocean Dome, located directly above this area.

The pressure on the Dome is slowly crushing down the ground floors of the Drone Anthopheridae Warrens, home to the digger BayBays. Meter by meter, annuation by annuation, the upper Hives will slowly crush down the weakest link in the structure, the point where the Hives connect to the Warrens.

The WC finds little problem with this, because they deem that the Warrens house the dregs of society. Most of the AGGIES originate in the Warrens, or they're devolved Bay Bays (from the Hives) who never manage to work their way back up to their home Hive.

Once we found out that this area was evacuated due to the toxin exposure level, we moved in. The WC hasn't given this structure another thought, so we feel reasonably safe here.

We move around a lot. We don't carry around that much gear, for obvious reasons, we have to travel light.

But, I do have to say with a modicum of pride, our personalized Eedia equipment is tut (our new members are required to give us all their worldly goods, so we get the most up to date Personalized Information Technology.)

That has to be enough to keep us going.

We clone their technology and change it to fit our special needs. We've used our micro-biotechnology to improve it and, occasionally, when one of our members turns on us and evolves back up into the Hives, we give them a reconstituted PIT module that infects their system with our tracer bug.

Oh yeah, we've got our fingers in every pie.

None of that is of much import right now.

The only real concern I have is for the firewalk that will take place this night cycle. For the first time in AGGIE records, we will subvert the loyalty of a Senselle, and an Odissi dancer, at that.

We'll put the first Senselle to the ultimate test. The one and only Damaria Rotiarte will walk the storm tonight.

As it happens, a lulu of a scattershower will be coming 'round in a matter of moments. I'll transmit the exact locale when the storm's touchdown time has been pinpointed.

Damaria must be ready to go instantly, on my command.

Yes, I know, she's brand new. She just came to us. It still amazes me.

Somehow, in all this world-saving and building-up-of-alliances, I forgot that I am also….a man.

Once I laid eyes on her, I helplessly watched as the delicious morsel invaded my very mind-consciousness. I remembered the moment with a jolt akin to physical pain.

She is entirely unlike others of her ilk.

Was the viewing arranged? A diabolical plot to entrap me? Why was I, of all people, privy to a private healing session of her sacred dancing?

Certainly, she was unaware that I was watching. It was a healing ritual and she was completely focused on her work. The ritual between one of the WC talking heads, Laird, and his Senselle consort, Randi, is yet another trick by the Senselles to sway a WC vote.

Of course the meshed couple was extremely attractive, but it was Damaria who instantly captivated me. I was hooked.

Damaria, whose eyes are hypnotic, filled with sexual promise. Damaria, whose motion begs me to taste her hips, exotic flesh smelling of musky perfume, enticing me to find out her sweaty secrets. I will not easily forget it.

I'm growing close to becoming obsessed by her, can't keep my mind on business. Always reliving that forbidden transmission in my mind, wishing she were dancing for me instead of them..

Then, she came to me.

She showed up at the last AGGIE meeting. And even though I know better – even though I know I should not leap too quickly upon an opportunity, it was too good to pass up.

She came here to offer herself up as a tool to help us bust open the Senselle organization. The temptation was too great to refuse.

Between the message and the messenger....ai! My Achin' Hard On!

The Senselles are the very heart of the evil triad who act to enslave us as a people. The PHDs are addicted to their Sensationalism, their egocentric emotions and feelings, all lewdly for sale.

And, the MagicoMathematicians are no better, simply keeping the machinery working, the metal monsters that keep us all INside. (Sometimes I feel as though I suffocate on this canned air, these metal walls.)

Before the Senselles, the Media Overlord Barons were the original masters over the viewership; hoarding all the wealth, making us weak and addicted and insulated so that we'll never evolve beyond being a civilization of watchers, instead of doers.

I'd observed what they had done to us and concluded that in a sense, they were right. Being fed a non-stop diet of mind candy, I realized we'd never become more as a species. I craved the flavor that they stole away from us, the originality and zest for life. They'd reduced us to merely surviving in order to consume more of their products.

Once all the basics are provided for you, and mere survival becomes the highest mark you can shoot for, it sucks all the life out of you. We became consumers for their whorish wares.

Naturally, when Damaria struts in here offering insider information on the Senselle organization, I could not refuse.

It was more than her body, more than her willing and sumptuous need to sexually satisfy me. That's what she sells to everybody else on this planet. Of course, in their commonness, they greedily suck it up from her.

While it's true that I want that from her, I want more.

I want her allegiance; I want to see her burn with the desire to help me build this new society.

Now that we've shared a communal jack-in (during the AGGIE MeetUp), I know more about her intentions.

She's lived in the middle of one of the most powerful groups on this planet for long enough to know that it's crippled. The Sisterhood of the Senselles is dying. It's rotting from within, and she plans to help me boldly slice off the infection.

She wants to bring a bright new reality into this world. Just like I do. Oh, I've been fooled before, but not this time. With her, I know her conviction is real.

She came to me.

She came to me and volunteered what everyone else on this planet must pay a princes' ransom for.

To me, she gave herself freely.

That alone means more to me than all the chedits in the world. At the same time, it's damn frustrating that she's put a hook into my heart. I know it's wrong to be vulnerable to her.

I can't afford to risk the lives of any of my small group. For making me wonder about her intentions, and for taking advantage of my personal weakness, I am going to test her.

Severely.

Without the endless cycles and quarters of teaching, of self-discipline, of truly testing her real motives, there's a good chance she won't survive the firewalk.

I've arranged for her to receive intensive training, and with only 20 hours of training, she'll be walking the fire, left to face the dreaded scattershowers, alone.

Death streaking down from an uncaring universe.

For all the prying eyes that will witness her trial as she faces her walk, I too, must appear to be uncaring.

I shouldn't even express what I'm feeling right now.

The firewalk begins.

* **

The Puzzle

A family
though complex, varied
and constantly changing
is still a puzzle to me.
And after 50 years
I finally realized why
mine is sadly
off kilter
while I am there.
I'm afraid to admit
that although to them
I am born,
they have decided their puzzle is
complete,
filled in corner to corner
and I am
the extra piece.
Perhaps I belong to another puzzle?

By Barrage Rhyzin (5037 C.E.)

Barrage Rhyzin, with one hell of an uncharacteristic check-in, being the last one that good ole Asna, my procreator, will receive from me. It's a one way communiqué, so I can be completely honest, for the very first time in this little charade we call life.

I'm tired of failing.

I fail and fail and fail.

Now, don't get me wrong, I'm told I'm excellent at being whatever it is that I was programmed to do, (which for some inexplicable reason has been kept from me.)

Most anyone, from the lowest BayBay to the highest PHD, from the Supernatural Senselles to the meanest vendor in the FLEE market, can get free World Council endorsed access to information about their own programming.

But not I.

Reason being.....I have absolutely no idea.

Apparently, my origins are to remain an enigma to everyone, including myself, so I don't have anything to lose by laying the naked truth on the line.

I'm not good at being a bad guy. Oh, I'm cool with violence. It's just a job. But that shouldn't mean I can't express other aspects of myself. It would be so much easier for the MOB if I could settle for doing only what is expected of me.

I'm to be satisfied with the position of second banana all the time; to be "second in command", "the old man's kid", "His right hand man", the "heir to the throne", but it's no pleasure cruise for me.

By Vee, who wants to rule over a bunch of fat, lazy, stupid, Media Overlord Barons? They've got the RQ percentages stacked in their favor, they own the second single-largest holding of WC stocks, everything's bought and paid for. Watching the same old tired triad of power players attempting to rule Terrax is crushingly boring for me; to the point that even their troubles are trite.

On one hand, we have the infamous Senselles – bringing us authentic "sensation"- all aglow with their alluring emotions (that end up jacking you into unreal realms of physical addiction.) They are creating a world filled with dull-jacks, no more than insatiable buyers for otherworldly whoresons.

On Terrax, it doesn't matter what you're selling. If you can create a big enough buying buzz, if you can offer them one new substance, then convince the viewership they simply can't do without it, you've cornered a market.

Personally, I find no adventure in living like this. There's no real experience to be had. Oh, you can purchase someone else's memory of an experience. You can purchase their hot, sweaty sex. You can purchase their mind bending drug addled trips.

But you can't purchase your own personal freedom.

That's the one thing you can't buy, despite what the self-proclaimed AGGIES believe.

From what I've seen of the vidiot Above-Grounders, I don't think any of them would be able to survive direct contact with the OUTside environment.

It's no place for the softheaded AGGIE BayBays who've been devolved from their cushy Hive cubbies. Even for the lowliest of their ilk (the few AGGIES who're used to the slight amounts of enviro-leakage from the Warrens), it's deadly OUT there.

I've gone firestorm chasing with some of the best personal protection and the thought of going OUTside without it sends chills through me. (I'm not sure if that's considered an emotion, but it's as close to being fearful as I can be, I suppose.)

My "greatest fear" hasn't been fear of people. It's the terror (and thrill) of flying through a scatterstorm while I'm chasing some errant WC proclaimed rebel.

I get a volt of energy, thinking about my own bloody death, alone, OUT in space.

You see, I've no sense of personal importance, due to my eclectic genetic 'gramming. To a certain extent, you could say I'm a prime example of genetic engineering gone awry, which can be expected in human gen-mod involving too many variables.

The mixture they threw together sparked something brand new.

Me. The Dark Horse.

A PHD who is able to enmesh myself, my personal energy signature, with pure electrical energy.

My ability must be from Nature, it certainly isn't due to nurture.

On Terrax, I've received no "nurturing", in a literal sense.

It all goes back to the age old question, "Do the ends justify the means?"

Should we, as a race, continue to exist, just in order to exist?

Without "live" interaction, Life is static. Most of the viewership of this planet has seen it all, and done very little.

Turning to begin my next assignment at hand, I reprogram my destination coordinates, checking the accuracy of my lons and lats.

As I pull away from the Crystal Planet in my newly retrofitted space bot, streaking toward Phaeron, I'm trying to figure out why the Boss wants me to follow Arpad and Frankie.

Inside my head, I don't think this is going to be as important to the MOB, or the Senselles and all their intrigues, as what happened to me at the AGGIE meeting. "Meshing" with a den of strangers, losing myself completely in an electric ecstasy, I felt closer to these PHDs than I have to anyone else in my entire life.

Weirdly enough, when I was hooked-in to everyone else's life energy, mingling with their personal disteldt force shields, I felt a rush of "oneness" I had never felt before.

PHD, TDM, dark entity, P-energy, I sense that we're all made from the same source material. I imagine that I might be a newly 'grammed genotype, created to help bring forth a new wave of humanity.

Why do I think this way?

It's about the technology, BayBay! Everyone knows a disteldt shield is placed around the meeting's parameters to keep intruders like myself out.

Only, once I was inside the meeting, sharing the communal jack-in, it was too late to deny me access. I was absorbed directly into their comport system, becoming energy itself. I transcended more than just my physical body. I became one with the essence of the circuitry.

It's a feeling that is almost impossible to explain. It's like experiencing 100% VR programming for the first time, except it's even more intense than jacking-into an entire Senselle episode.

(Most of you BayBays have never jacked-into that much Senselle 'gramming at one time, cuz you can't afford it. Due to my high MOB ranking, and the fact that I have more chedits than Vee herself, I do know the difference.)

But none of that can begin to compare to the feeling of complete abandon I had while electronically conjoining with the AGGIE group. During the meshing, every molecule of my being was disassociated.

I was free.

I was everybody. I was one of them. I was all of them. I was none of them.

At the same time, I even felt more individually myself than I'd ever felt before.

Of course, in order to remain undetected, I couldn't maintain total dispersion for long. Or, maybe I'm kidding myself. Maybe I could've tried to maintain that new consciousness for longer. But I tweaked out.

Did I reassemble just for the sake of not being caught? Did I dare to go far enough, or did I back down in fear?

The new sensation was almost too good, too exciting, so I dumped myself unceremoniously back into my ridiculously fragile, yet plodding exterior body. At first my body felt unbearably heavy, hard to move and manage at first. I felt clumsy as I retrieved my jack-in mod and prepared to leave the meeting.

I began to feel that having a physical body, is to be in a kind of bondage, no matter how fragile it may seem.

We PHD's have lived in a gravity starved atmosphere for so long enough that our bodies have become dangerously attenuated, stretched thin and rather weak.

We learned to cover up this fragility by projecting virtual holograms onto our body's surface, in an attempt to look more fierce.

The practice became fashion, and the rage hasn't abated since then. At this point in time, our natural bodies haven't been exposed to natural light for eons.

Thus, being unable to absorb real star energy directly, we lack the needed UV rays that would keep our natural bodies looking vibrant. The result, our real skin color is gray and ashy (the color that results from the genetic mixture of all the races. I'm not even sure it's a real "color".)

Color is a fairly useless demographic anyway. Some PHD's claim we're blue. Some say we're gray. Most PHD's wouldn't dream of looking at their "real" body, because it's too much a shock to see how degraded they've become. So the "color" question is now moot.

Although I've no distinctive features on my "real" body, I tire of keeping the holograms on. I've come to accept the way I really look, so OUT here in space, I drop the holograms, at least while in orbital transit.

As for me, I'm riding the wake of these two inconsequential PHDs, Arpad Reitan, the abominable HORSE, and his filly, Frankie.

I'm kinda wondering what the hell Arpad sees in her. I mean, who'd be interested in such a tepid little waif? Frankie's not anything like the lascivious Senselle, Randi.

Frankie's a PHD, she's never "lived" any kind of intense experience. She doesn't carry herself like Randi, a Goddess who could do anything, at will.

Frankie sits all scrunched down, and she walks hunched over, like she's trying to hide from the world.

Lah, I don't get it!

They're freaks, that's all I know. And then there's me … the little tail end of a freak. The big bad wolf who would rather play with grandma then blow down the house and eat the pigs.

Olde Earth mythology aside, I must say, the farther away I get from Terrax, the better I feel. I experiencing an incredible "lightness of being", and it's solely due to the fact that I'm communicating this message, not through the data, but as the data.

I am ones and zeros.

I am the light that's been tripped – fantastically.

I am dispersion and disimaging. I can choose to recreate myself.

Although technically, right now, I can only zap myself back into the same body from whence I came.

Hmmmmmmmmm, that's another thought.

What would it feel like to mesh with another PHD's body, becoming them for a brief time, using their physical being for my own intents and purposes?

Right now, for the very first time in my existence, I'm just happy to be who I am.

I'm no longer simply "the great fear's shadow", Asna's henchman; his doer of evil deeds, large and small.

In future, I may even take on a new name, but for now I will remain Barrage.

I'm facing an electrifying new world, hurtling toward my destiny, a new life that will be more real to me than all the other claptrap I've been offered up to this point.

* * *

Phantom Energy-Matters

We are the Unnamed, wholly comprised of phantom energy that has been coalescing for millions and millions of eons.

Time, spanning in all directions – space, spinning through all dimensions – all contain a modicum of dark matter, as well as particles and waves.

We were without thought, We had existed without intention, until by chance we meshed with a quixotic mixture of toxic poisons and antimatter.

Antimatter formed when the explosion from Olde Earth's dying sun contacted the coalescing field of phantom energy that had been collecting around its planetary body, it caused the Earth's energy field to implode.

With disastrous repercussions.

The impact from the implosion had the power to shatter the Earth, however, at the moment of implosion, the Earth was regurgitating back a toxic cocktail of chemical poisons and nuclear energy that had been steadily injected into its surface for the last few centuries.

The last explosion of the tortured Earth connected with the inrushing phantom energy. At the precise moment these energies collided, the very fabric of Time ripped apart.

The Rips in Time affected neither the location of the Earth nor the formation of the newly charged P-energy field.

Once the Time Rips occurred it sparked the antimatter that mixed with our energy, which invested Us with a new characteristic. We "became" intention itself.

Our first "intention" was pure action; because that's the opposite of what P-energy had always been. We had always been inert.

Inertia is not felt, except in its absence. Once we absorbed the antimatter that caused the Time Rips, the P-energy was then negatively charged. We became the Dark Ones.

This created intent – the very opposite of the void- the very opposite of our previously random, meaningless patterns of energy. Now We have a signature all our own.

We began wandering time/space to find a physical host in which we could manifest our next "intention"; to act.

Satisfied with our findings, our intentions were transferred through, and absorbed by, the energy/entities enslaved on Rehabco, Phaeron.

The Retainees of Phaeron are consigned to live there until their natural life span ran out. They are left to dwell forever encased in a cold, featureless automated building.

After the initial energy transfer, it only took a few annuations for the residents to figure out how to "become" pure energy with Us. We discovered they were able to communally mingle by "jacking-in" together and jumping the energy grid barriers, leaving their bodies temporarily. They enabled the doppelgangers (the bodies they left behind in their cells) to continue to perform their daily tasks.

Their absences went undetected.

Any Guardian reviewing their Retainees data feeds would observe a "normal" readout from the data streams, and would assume that living, breathing PHDs were inhabiting their cells.

For all intents and purposes, the Retainees appeared to be doing what was expected of them, patiently waiting for their internal clock to rundown, like good little lambs to the slaughter.

Instead of passively waiting for death to relieve their ennui, they'd begun playing energy games, similar to what Barrage had discovered back on Terrax at the AGGIE meeting.

Once they discovered that they were part P-energy, and could potentially become pure energy, they practiced melting into a communal jack-in within the confinement of their localized disteldt shields. They call themselves "The Mind Gamers".

Barrage hadn't had time to learn to mesh into a completely open system, but The Mind Gamers on Phaeron had nothing but time on their hands. Long ago they figured out how to project themselves holographically from one place to another.

Soon tired of playing confined within the limitations of their physical bodies, they realized they could shed them as easily as a soiled change of clothing.

They learned to become pure energy, and jacked-in to the local Phaeron grid.

Since no PHD was denied entertainment, even those damned to live on Phaeron, the Eedia Net had an old broadcasting system working on Phaeron. That's the very same equipment the Mind Gamers used to transmit the signals to the ROV on Terrax.

And it responded.

That's how the ROV on Terrax received the unauthorized programming for that outgoing communiqué.

The Mind Gamers never tired of jacking-in to their local energy grid, experiencing their communal energy commingling, and the wonderful release that comes with complete dissipation of self/ego.

It is a gift, to be relieved of the weighty self/ego. The incredible feeling of joy, of blending into other life forms that aren't encased in a particular ego, with all its limitations, can't be duplicated by any VR 'gramme on the market.

The Mind Gamers savored this new way of "being", of enmeshing oneself into the group energy, ever flowing and changing, simultaneously creating new entities and patterns.

Soon The Mind Gamers found that they were seldom completely returning to their physical bodies. They were able to put their body on automatic pilot, programming them to have the appropriate actions and reactions to their Guardians.

Their "evil twin" made sure that all their biomed readings were accurate. These canny Mind Gamers would then leave their doppelganger in their cell to do the daily drudgery of editing, purchasing, and supplying E-media, "the entertainment for the mindless masses", to the other bored Retainees of Rehabco, Phaeron.

The head honcho of the Mind Gamers is called, "Vortex."

Vortex assembled Rehabco's mentally strongest, most physically fit Retainees, and asked them to focus their attention on a single purpose.

Similar to the way the dispersed light in a room can be focused into a single laser beam, he collected their communal energy together to gain the power they needed to jump the (supposedly impregnable) firewall between the Phaeron and Terrax satellite systems.

Vortex chose to transmit to Terrax primarily because their PHD's energy signatures, once synced with the Phaeron signatures, would be similar enough to allow them to easily infiltrate their Vidnet.

Vortex despised Terraxians, considered those PHDs belonging to the past, to a defunct and useless form of civilization.

They're encased in weak physical bodies; attenuated and thin, their outer skin almost colorless, and without any distinguishing features, his haunted thoughts intoned.

Even with all their holographic projections to color themselves up, they're no more than consumers and disposers of electronic media.

They're nothing more than a bunch of drones who've been

programmed to sit and watch their 288 mini-screens all day, imbibing in e-juice, and living vicariously through their precious Senselles in a miasma of virtual reality 'gramming.

Beings that aren't producing anything, have no original thoughts, nor the ability to create, they're no better than puppets, the virtual shadows on the cave wall.

Besides the fact that they're wasting their lives away, they deserve to be our lab rats because "turn about is fair play".

They sent us here to Phaeron, so it is fitting and right that they will be our first experimental subjects.

We began our program of dis-ruption by communicating with that ROV on Terrax, and then we infected their local grid when they tried to deprogram it.

Eventually, the garbled outgoing transmissions started to wreak havoc with the power triad back on Terrax. Since nobody there trusts each other (and being allied in any way was completely out of the question), they were easy pickings, really.

We played Asna's overweening ego against Wad's complete self absorption, and figured they would never come face to face to compare notes. They'll never be honest enough with each other to admit what they were holding back, so even if they do communicate face to face, they'll share no critical intel.

This forces them to be fragmented. They still hold on to this unworkable concept of "ownership", and a kind of "me vs. them" mentality. They'll likely never be able to solve the problem alone, and won't work together.

Therefore, they can't possibly solve it.

I know that if we continue practice-jumping the firewall and transmitting these mysteriously encoded messages; our "game" will serve a two pronged purpose.

First of all, we The Mind Gamers will get strong enough to transmit outward to other galaxies.

Our first attempt to transmit outside our LAN was chaotic. The signals didn't actually hit Earth, but we did manage to install the interrupt programming into one of their ROVs. We had it ram into their Ocean Dome in such a way that it would cause extensive damage.

As anticipated, it caused the power factions to personally interact with one another.

It's pathetically funny to watch them try to figure out what to do.

The MagicoMathematicians are static individualists, each trying to drain their brain to break that ROV's outgoing code. Too bad they're wasting their time. It's gibberish, meant only to distract them while we are infecting their system. The only way for them to survive the attack, would be to unify their internal factions and realize that the attack is coming from an outside source.

They may be smart, but they don't understand what's important.

Same thing with the MOBsters. They're so used to corruption and betrayal that they figured this was a double cross by either the Senselles or the MMs. Asna may have even wondered if Grandpa and his old cronies could've had something to do with the ROV transmissions.

All the MOB knows is that since *they* can't be trusted, they think *no one else* can be trusted either.

Old Fool. Nobody saw the obvious.

The PHDs have absolutely no motive for destroying an Ocean Dome. The Domes contain all the most dangerous environmental toxins, protecting all the viewership, as well as function as a crucial power system for Terrax.

Can't the Triad see how ridiculous it would be for any of them to be involved with this? If they weren't so frantic to blame each other, they'd see that it's impossible for the code coming out of that ROV to be either MOB or Senselle generated.

The Senselle gen-ennges, for their part, had no idea whatsoever what the code meant, or why it was being beamed back to Phaeron.

Shaking his head, Vortex thought it was certain this mysterious event would cause a visitation from the good ole home planet. He thought they'd at least send out an information gathering soiree to Phaeron to come see what the hell was going on out here.

But…nothing.

Yeah, Terrax performed a few cursory information sweeps by remote, but no one showed up in person.

That will prove to be an irreversible error.

Now, we're capable of sending part of "our" energy to Terrax undetected, while another part of Us remain here. After we infect the computer systems of Terrax, we'll be able to control the life support systems of the entire planet.

We don't know of what value this might be to Us. The undertaking is a vast experiment.

At this stage, we feel like a patient who's slowly recovering from a long-term illness. Suddenly, we discover that we've been too careful in our attempts at normal activity. We find that we're stronger than we'd ever imagined, ready to flex our proverbial muscles.

We'll not be happy to merely crawl or walk, but will stride, then quickly, run. We begin to feel the strength flowing through us.

We intend to be, once more, the deciders of our own fate, not just exiled prisoners, left to rot until our timeline winds down.

Beware, we are The Mind Gamers of Phaeron come upon you.

6

The Dark Ones

<u>Too Cute to kill</u>

You're at my window, baby
throwin rocks at 2:00 a.m.
You break my dreams,
you trash my sleep,
but then I let you in
because you're...
too cute to kill
too cute to kill
too cute to kill
too cute to kill.

(song by Kayla Raconteur)

This is Auntie Vee, dictating my daily Senselle gram, and then it'll be off to work at the OTEC plant where I'm conducting additional safety inspections this night-cycle, due to the increased toxic exposure from that dratted ROV.

To kill some downtime on the Sextel Network gramming board, I'll leave you with a little vignette of everyday life back on Olde Earth I like to call...............

The veterinarian's visit

It's really strange how my furkids know when I'm taking them to the vet. Maybe I exude a certain telltale scent when I contemplate sneaking up on them just long enough to get them into that cat carrier. Or in Melanie's case, getting her out of the house and into the car before she figures out what's going on.

The morning that Melanie, our Stabyhoun dog, is scheduled to go to the vet's I don't deviate from our usual routine, so as not to tip her off.

I arise at the gawd-awful time of 4:30 a.m. (that's when the cats want to be fed) stumbling into the kitchen, eyes half closed, to make coffee. I give Melanie a few kibbles of dry dog food, as I putter about mixing half a can of cat food with half a cup of their dry food, dispensing it into the six containers for our pride of wild and semi-domesticated cats.

While the coffee's percolating I watch music videos and dance to a tune if the mood catches me, or open the drapes and simply watch as the light comes into the room from the east as the sun begins to rise, filling it with mellow morning light.

Melanie comes to sit at my feet begging for more breakfast, as the cats, now satiated start to bug me to be let outside. Sometimes Melanie will go outside with them, getting an early start on her daily doggy duty of barking at the squirrels running around and around the bole of the huge oak tree out front.

As soon as she starts barking, I bring her back inside because I don't want to disturb our neighbors at this time of the morning.

After Wade returns from taking Melanie for her morning walk, even though our routine has remained exactly the same as every other morning, she begins to get antsy, as though she knows what's coming next.

This morning, as always, Wade dashes off to work in his truck, but instead of letting Melanie back inside for her usual morning nap, I take the leash and load her up into my car. I've already prepared the rear seat with the special plastic and cotton liner, so that if she has an "accident" she won't mess up my car.

I strap her into a doggie safety vest (a harness that straps around her chest and is attached to the car's seat belt from the back of the vest), so that if I have to break suddenly, she won't come flying forward and go through the windshield.

Unlike other dogs, she's not that excited about "going for a ride in the car".

She gives me that piteous look of hers, as if to ask "What on earth did I do to deserve such punishment?"

My adorable Stabyhoun pooch is a blue-merle color; her coat mottled black and white, with thick curly fur on her hind quarters, and long silky hair on her belly and legs. A medium-sized dog, weighing in at about 35 to 36 lbs., her short legs make her look rather stout. Her head doesn't come up much higher than my knees.

Her face has a distinctive white heart-shape framing her eyes and curving around her muzzle, creating an eternally inquisitive look on her face. She does the whole cutesy-pie dog trick of tilting her head to one side, and raising one eyebrow as if to say, "Aren't I cute?"

Yeah, you're cute, but you're still going to the vet whether you like it or not.

She's got this gorgeous full tail, resembling an old fashioned plume pen, which sticks up jauntily as she vigorously wags it, waving her happiness for all to see. Her dense undercoat and long hair on the back of her front legs, withers and underbelly keep her way too hot in the summertime, so at present she has her summer cut which makes her look years younger.

Originally bred in the Netherlands (that's why her breed is spelled Stabyhoun, without the "d" at the end of the word), she is supremely suited for snow country, and is listed as a rare breed.

She is the smartest thing too. A herding dog, her genetic makeup gives her the intelligence to make decisions while on her own, far away from her master as she tends and herds sheep. This makes her independent, yet loyal to the family group, and a most loving companion.

With her deep chest, she isn't afraid of using her commanding voice to get what she wants. Although she's a small dog, she uses that big voice to insist on being heard when she wants something.

As we take off for the vet with her secured in the back seat, I have her lie down so she doesn't lose her balance and fall forward off the seat.

Coaxing her to relax on the way to the vet by playing music, I start singing along with my new Lenny Kravitz album (banging on the steering wheel in time with the drummer when I really get into it.)

By this time, Melanie's getting anxious, looking out the window and wondering what's going on. Occasionally she'll let out a small yip, but doesn't let go with her full throated bark, thank goodness. (In this small space it would be deafening.)

As soon as we get to the vet's office, she very much wants to get out of the car. I get out and go around to her side to help her disembark, because she can hurt herself trying to jump down from that height on those short stubby legs of hers. It's a bit awkward, as I simultaneously hold her leash in one hand while keeping my bag slung over the other shoulder, but I manage to help her out of the car, and off we go.

She's becoming more animated as we head for the front door of the vet's office, with her tracking every scent in the parking lot, sniffing and walking around and around in tight circles. She can smell all the other dogs and cats that have been out here before us.

As I approach the door of the vet's office, I can already hear the din from within. This is a busy clinic because they don't require appointments for regular services like your pet's annual shots, and they even take some emergency walk-ins. They also board your pets while you're on vacation, making it less stressful on the owner because if anything goes wrong while you're away, your pets have their own veterinarian there to treat them.

A friend of mine coined the phrase "furkids". It perfectly describes what it's like to own a pet, because it really is like having your own kids (only they're the four legged variety.) This works for everybody all around because if you call your pets your children, some people go absolutely crazy. They think that you're saying animals are as important as people, personally I can't see anything wrong with that. To me, my furkids are as important to me as other people.

At times, they treat me better than people because they give me unconditional love. My furkids are always glad to see me, and not just because I feed them, even though these same people will make the comment "sure they're glad to see you. To them as you look like a huge walking can of cat food!"

I know that simply isn't always the case, because even after they're replete with food, they still come around for companionship and love. They're here for me whether I'm sad, or a good mood, and to them I'll always be "mom".

They remain playful even into old age, and I say these things with confidence because all my adult life I've had pets around me.

But I digress.

Once inside the clinic, Melanie is quite a character. She needs to be the dominant dog, and even though she's small, she thinks that she's big.

As we stand in line to sign in, the doggie encounters begin, and although each dog must be on a leash and each cat in its carrier; it's still general pandemonium, what with the dogs interacting, the cats becoming more uncomfortable by the second and the children zooming around the waiting room.

The waiting room has a distinct reek to it. I mean, they keep the place clean and all, but after years and years and years of animals having accidents on the floor, believe me you can still smell the faint odor of animal pee mixed with the cleaning agent they use.

The children who come with their parents are joyfully running around shrieking, trying to pet as many of the dogs as they can, while the front office attendants call out the next room assignment. Some of the dogs are fearful, shivering and shaking like they're gonna fall apart; others are simply curious and excited, dashing from one dog to the next, sniffing and growling softly.

Everyone is trying to figure out the pecking order.

Of course, the cats are pretty nervous, with all this dog energy crowding around them. They're either hissing; scrunched deep into the back of their cat carriers, or worse, freaking out as they smash themselves up against the cage over and over again in a futile attempt to escape.

Melanie will usually try and make friends with the people first, then get to know the dogs. She usually minds me well enough, but these are unusual circumstances, being around so many other dogs. I let her go and sniff the other dogs, but she's usually more interested in the kids.

It's pretty hilarious. I've seen her try and subdue the big dogs, ones who are three times her size, because she has to show everybody who's boss. And, she has to get in the last word.

Anyway, we finally get into our room (room #4 for today), and I close the door to the hallway to keep her from getting too distracted by the other dogs walking to their rooms.

The front office staff made notations on Melly's chart to let the doctor know what we're here for (whether it's just a routine visit for shots or if there's a particular problem.) This way the doctor knows what to look for once she arrives and can scan Melanie's medical history before walking into the exam room.

Since Melanie is on the ground, I need help to hustle her up onto the examination table. I have the doctor help me, because I'm only 5'1" and the table is about chest high off the ground. We lift her up onto the cold, sterile aluminum table so the doctor can get a good look at her.

She's behaving herself, partly because she's a good-natured dog, but also because the cold feeling of metal on her feet is part of the control process. Plus, she's in an unfamiliar place with a strange handler. She's not used to strangers coming up to her and handling her with this much familiarity, which puts her off-guard and gives the vet an advantage.

They take her temperature rectally, which has her a little uncomfortable. She looks back at the doctor, as if to ask "What the heck are you doing?"

Dr. L. combs through her fur to see if she has any fleas, looks into her eyes, ears and mouth (to check to she needs a dental cleaning.) They already weighed her up front in the waiting room, and I usually get balled out because she's too fat.

The thing is, the dog is low slung; she's low to the ground with itty-bitty legs, excellent if you're bred to herd sheep. In the coldest times of the year, she gets to lookin somewhat wolf-like; with her thick furry coat and long tongued doggie grin.

Her plumed tail sticking jauntily in the air, she'll dash after anything that will give chase. She can be rather intimidating, snarling and nipping; but her teeth aren't made for shredding. She has short teeth for nipping at the heels of sheep, which she can control by herding them this way or that. That's the way she operates.

She's never bitten anyone, that's not in her character, but whatever she's chasing at the moment sure doesn't know that!

She's a super intelligent, loving, independent being with her own personality; a pain in the neck... and love of our lives.

My husband and I have never had children together, but we've always had furkids. However, Melanie was our first dog.

Once she gets her shots and medicines (if needed), Dr. L. says "See you next year".

And I know that our annual visit is over.

Whew. We get in line to pay and then rush home to roll joyously on the carpeting together.

***

Auntie Vee ended her 'gramming mode and quickly scanned the open channels for an encoded email. Switching to the local OTEC grid, she prepared to continue her inspections for the day. For a moment, Auntie Vee couldn't help worrying about what Asna may have in store for Wad at their impending face-to-face meeting.

Of course, she trusted that Wad was strong enough and smart enough to meet "the Great Fear" without personally being harmed. That wasn't what she was worried about.

It was what she felt closing in all around her that had her jumpy.

She could sense big changes about to happen, and whenever that busted loose, she wanted Wad by her side. No matter what was coming down the pike, she'd feel better with him here.

She couldn't bear to think about what might happen to this world once they were gone, but she couldn't worry about that right now. She had to do what her heart told her was right.

And right now, her heart was telling her it would be okay to wait to begin her data recording for at least a few more nanos, hoping Wad would peep her.

Meantime, it wouldn't hurt to listen in on Cami's latest escapade, it sounded like one she hadn't heard before.

As the scene begins to unfold, Auntie Vee allowed her problems to be relegated to the back-burner of her mind, while Cami spun yet another one of her fantastic yarns of yesteryear.

* * *

"Happiness is NOT a Warm Gun"
-Cami's Latest Escapade

Speed and I are living together in Orange County. It doesn't really matter where. All small towns surrounding big cities are the same, that's why they're called "the burbs".

We work well living together, he and I, primarily because we aren't "together". I found us a two bedroom house to rent; it's on a street that takes a Y around each side of the house, thus the cheap rent. We don't care much about the street noise, we're prepared to make our fair share and this means there'll be no nosy neighbors around to complain about it.

Part of the street swoops past the house on one side, and then leads to a tiny alley that dead-ends into our driveway. Very private, it makes it easy to swing into the garage from the street. Speed is all for this because he can put his Yamaha 450 in the garage and walk straight into the house from there. Works for me because I have my own bedroom and blessed privacy.

When you step into the back door you can see Speed's bedroom straight ahead, to the left is the bathroom and down the hall on the right is my bedroom, tucked away in the back corner of the house.

From my bedroom, I can enter the tiny kitchen, which then opens into the living room and the front of the house.

The living room is the party area, which doubles as the guest bedroom, and you can believe we'll have lots of hard-partying friends who'll crash here overnight right on that hideous green couch of ours.

I work as a chiropractic assistant at a clinic near here, run by a megalomaniac chiropractor, and Speed is driving truck for a local company, delivering booze.

We like to ride the bike and party on the weekends (not necessarily with other bikers). We figure we can get into enough trouble on our own, we don't want to have to run with a crowd and act a certain way.

Sometimes we'll head up the coast on PCH to Zuma Beach, or we'll fly down the freeway to South San Diego to visit Bear and his drinkin' buddies who live near the naval base. Man, can those boys drink!

It's been a hard work week for both of us, so we decide to kick back at the pad this weekend. Instead of goin riding, we decide to score some dope from Ricky, our neighborhood connection, and hang out at the house. No big plans here.

Ricky runs a dope house, meaning, the local cops know about his "business", but they don't pop him, because he's small time. It's not like he deals heroin or anything like that, he just sells dime bags of weed and some coke now and then.

He doesn't deal the heavy drugs the cops watch out for, so they let him deal his pot without hassling him, as they have bigger fish to fry.

We feel pretty safe going over to Ricky's anytime, so Speed and I jump on the bike and cruise on over. No need to call ahead or make an appointment, Ricky's almost always home.

Since he sits around his house a lot, he's picked up different hobbies to help pass the time while waiting for his customers. At the moment his obsession is with those toy racecar sets, like the ones your kid brother used to get for Christmas.

They're pretty fun. I'll take my turn racing every once in awhile, but the guys are much more serious about winning.

Ricky set up the tracks in his living room, and people who are waiting to score can race two at a time, each driver choosing their lucky car color. With remote in hand, off they go!

Every time I try and race the darn things, mine tumbles off the track as it makes the first turn and my race is quickly over.

Damn! Happens every time, so I lose interest and stop playing.

Anyway, there are two guys here already racing when Speed and I arrive, so we go to hang out in Ricky's kitchenette, which has a backgammon board set up ready to play.

That's Ricky's other pastime, and something for his other customers to enjoy, other than watching TV or racing, while we wait to score.

Ricky's off in his combination bathroom/bedroom (which is where he keeps the dope) with the door locked, weighing out the orders. He goes in there with your money and comes out with your stash. If you want, you can hang out with him for a while, or just jam, if you have some place to be.

Ricky doesn't take it personally if you don't hang around, he knows it goes with the territory, cuz after all, he is your dope dealer and not your best friend.

Most people don't hang around for long, unless they want to smoke up his pot, but that's nothing to Ricky. He makes his real money selling a little coke on the side. As far as he's concerned, if you'll hang out and keep him company, he'll let you smoke his weed all day.

So here we are; Speed and I playing backgammon in the kitchenette, and by now, there are four guys waiting in the living room, two racing the cars and two waiting their turn to place their orders.

The two guys racing assure us that Ricky's probably ready to come out with the first round of orders any minute now, so we kick back and enjoy our round of backgammon.

We've only been here a few minutes when suddenly we hear a tremendous BOOM, and the front door comes flying off its hinges, smashing heavily down on the living room rug, and the hapless racecar set.

Pieces of plastic and little cars go flying in all directions as we instinctively freeze where we are.

The front door's just been kicked down!

(The thing didn't offer much resistance, apparently it's not the first time it's been busted down.)

It comes down like a tree, making the apartment momentarily rock with the force. I hear this thunderous thud, and think "thank God nobody was sitting in front of that door!"

Everyone in the room is in shock. Time stands still for a precious second.

Too soon, I'm aware of a gun pointed at my face, at close range. The cold metal shocks my skin as it touches my temple.

"Everybody get on the floor, face down, or I'll blow your fuckin heads off!" a man yells unnecessarily at us.

I'm absolutely unable to move once I fling myself face down on the floor, the cold kitchen tiles feeling cool to my fevered brow. I wish I could disappear, but I'm too terrified of that gun to move.

You see, from the safe distance of a locked gun cabinet, a gun's barrel may appear to be a finger-tip sized hole. But when the thing is pointed at your head, it's the biggest, darkest maw you've ever seen in your life. With a sick vertigo, I feel like I'm falling into the barrel of his gun.

And the voice behind that darkness? A harsh noise without a face, cold and uncaring.

I swear I don't even know what his face looks like, that's how terrified I am. Instantaneously, I want to make myself as small and unnoticeable as possible (I'm painfully aware that I'm the only woman in the room.)

I don't want to get raped.

After the first guy through the door points the gun at us, we leap to the floor without a single protest. From the sound of it, the second guy's moving swiftly from person to person, rousting us for wallets or money, anything he deems valuable. He shoves the guys around pretty roughly, challenging them to give him flack.

The third man is the silent lookout, I don't hear him speak at all.

Three guys against six people, you'd think we could overpower them.

But, one of them has a gun.

Only one needs to.

After securing the doorway, the third robber stomps in, nearly trampling the bodies cowering on the floor, and rips the phone from the wall. Then he goes silently back to his post, nervously shifting from foot to foot, anxious to be gone.

Lucky for us, Ricky's thinking quickly. He's been in the business of dealing long enough to know that depending on what he does next, he can either save our lives or we'll all end up dead.

Ricky comes out of the bathroom quietly, and closes the self-locking door behind him.

"Hey, I'll give you everything I have" he says clearly and calmly. He immediately turns out his pockets, revealing several little bags of drugs as they fall to the floor. He stoops to pick them up and extends his hand to the goon nearest him.

There are four bags of measured out dope, one mixed bag of multi-colored pills and the money we'd just all given him to score.

The guy with the gun looks skeptical, like Ricky's holding out on him. All three of the robbers are breathing hard, getting nervous standing around an apartment with a gaping hole where the door had been. This was takin too long!

They'd already gotten our wallets, all of Ricky's money and the drugs, but for a moment they hesitate.

The deal is about to go very wrong.

"Hey, look, I got this coke on me too", Ricky pipes up as he casually moves toward the still blaring TV, left ignored during the entire robbery.

There on top of the TV is a cat-shaped ceramic potpourri jar. He empties the jar as if in slo-mo, and out fall six tiny dime bags of coke.

The leader of the pack snaps them up, and motions his boys out the door with the gun.

"We've got a man with a gun covering you from across the street. If anyone so much as pokes their head out that door to get help, he'll blow your fuckin head off!"

And with that, they speed off, tires screeching.

Believe me, we're all too stunned to even think about calling the police. Besides, they'd ripped the phone out of the wall.

We're all talking endlessly about it afterward, like we're on a time-delay tape loop and can't stop hitting rewind. It's so unreal. Like, after you've gotten off a long plane ride and can't quite hear properly, everything seems peculiar and unfamiliar.

I mean, everything in your world changes the instant you go from sitting around doing what you normally do, to having a door fly off its hinges and come crashing down at your toes.

It's an experience I'll never forget, for good or ill.

The funny thing is, even though it was terrifying and all, we still score dope from Ricky. (He got a brand new door put in after that last robbery.)

I guess I've been lucky.

In all the years of taking dope, and dealing with the drug world, it's the only time anyone's ever put a gun to my face.

* * *

Auntie Vee, deciding that she'd waited as long as she safely could for Wad to peep her, continued her inspection rounds. She efficiently "cleared" each info-data field of her portable percomp as she "walked the line".

Chuckling softly to herself at Cami's endless escapades, she knew one thing for sure. No matter who she met in her near future, they weren't likely to be as interesting, or as adventurous as her rough and ready Senselle Sisters!

Good Byes Come in All Shapes and Sizes

Damaria took her time languorously waking up in the arms of her lover, Pavarti.

Making love to him is so sweet, so fulfilling. He takes his time. He teases. He makes sure that no matter what, I have my pleasure. And he allows me to pleasure him as well.

You know the drill, all the romance novels in the world have made all the details readily available. I share it all with my viewership, partly because it's nothing new, and partly because I don't have anything else to talk about. That's why I call him "my lover" in generic terms, rather than "my beloved", which would be like claiming him for myself.

I do not belong to myself, so I can't give myself away, and I definitely do not believe in "love", so there's no sense leading him astray.

I guess you have to understand my story from the beginning to understand the depths of my mistrust of people, especially those who call themselves Senselles. We simply come from a time past, and are certainly no better than the poor sods we call PHDs.

The Vidiots, Talking Heads and Posey Posers have already told the crude bare facts of my past. But they can't possibly know what it's like to be so poor that you lived on the scraps of leftovers from others, digging into trash cans for half eaten meals that were tossed away, living in someone's garage and peeing into an empty milk container because there was no plumbing in the garage where you sleep.

My folks didn't work much. My mom sat around on this ghastly tan striped barca-lounger, sucking down expired chocolates and ordering stuff out of magazines that we didn't need. She would fill our lousy little apartment, or worse, our temporary hotel room, with all these cheap knick-knacks and googas.

It drove me crazy because we didn't even have enough food to eat and here she is wasting our money on all this crap. And then there's my dad, who sits for hours doing nothing, seemingly just quietly desperate, until he'd lose his temper and start screaming at her about how much she spent on shit we didn't need. A real circus.

I remember one time she maxed out three credit cards buying these sets of ceramic gnomes to put in a yard we didn't have, and then it was angels, in every size and shape taking up every square inch of flat space in the apartment. She bought angel salt and pepper shakers, angel wall hangings, hanging mobiles, wind chimes, angel dish sets and mugs. My brother and I drew the line at wearing angel T-shirts.

As a kid I was always hungry because we never had anything to eat in the goddamn house except for cereal and sour milk. My dad tried to work but he wasn't really good at anything, so he wouldn't stick to any one job for very long. He acted like every new job was going to be a new try, and a brand new beginning, but his depression was quiet and deep, so he drowned himself in alcohol to numb the pain of his failures.

He would play all these little games to try and hide it from my mom, but didn't really bother to try and hide it from my brother and me, which made no sense to me. It was like my brother and I were the adults and they were the kids.

He drank every night so he couldn't always get up in the morning, and spent too many days callin in sick. Then when he did show up to work he was unshaven, rumpled and reeking of alcohol. He tended to get fired a lot. Which caused us to move around, we never really had a stable place to live, or enough to eat.

I'm not trying to start a pity party here, I'm just saying we were always scrambling to keep a roof over our heads. I know what it's like to have to dig through trash cans to find your next meal. It always astounded me, as we went "dumpster diving", what people are willing to throw away as garbage. We found clothes and shoes, old chairs, and all kinds of used TV's and telephones and stuff. Once I even found a perfectly good computer. At least it was most of a computer, but as usual we didn't have the money to complete the set, and it ended up being thrown out again.

We'd raid the curbside for bigger furniture like sofas, or even a dilapidated refrigerator or washer. It was all serviceable stuff, absolutely nothing wrong with it, and I can't understand why people just tossed it out.

Even though it was the neighbors' cast off garbage, I still felt like a thief stealing the second-hand junk. It made me and my brother angry and ashamed of our parents.

He and I had to work so hard just to look semi-normal, wearing clothes from thrift stores or filched directly off clotheslines if we were desperate for newer looking clothes.

So you can see how bitter I was (from having to live off people's charity), when this do-gooder lady, Cami, offered me a job I wasn't qualified for. She paid me a decent wage for once in my life, and without asking for one damn thing in return, she took me from a know-nothing retail clerk without a future to having some basic bookkeeping skills.

She gave me money to shop for some somber business clothing, and even bought a bed for me when she found out I was sleeping on the floor of my apartment.

But I began to hate her rules, her discipline, the ghastly serious dress code, the upbeat "pep talks" at the staff meetings, and her incessantly optimistic morning persona.

At first I was doing okay working there. When I finally dug myself out of debt, I began to pursue the ancient art of Odissi dancing, the sacred dance of India. It began to mean more to me than just a great way to lose weight; I started to come into a sense of myself.

Instead of feeling fat and bloated all the time, I learned to appreciate what I could do with my curvaceous body. I felt sexy.

Oh, I've always gotten a lot of attention from men (my big double DDs ensure that), and I can easily return the attention without any internal feelings at all. Mainly because my uncle had been fucking me since I was a kid (he'd started abusing me when I was only eight.)

Yeah, the asshole (my father's brother) was staying with us, when one night he came into my bedroom when I was sleeping and stuck his dick in my mouth. Obviously, I couldn't scream because he'd grabbed my hair and held my head in place as he slowly slid his dick down my throat. I could feel the hair on his balls tickling my nose, and I gagged at the smell as he worked his dick quickly in and out of my mouth.

It was over in a few minutes, and afterward, he leaned down to my teary face and whispered "you know you loved it you little bitch. Tell anyone and I'll only hurt you worse next time."

That was the beginning of the six years he sexually abused me. The next morning, I found out exactly how much "protection" I'd get from my parents. I told my mom, and she slapped my face for "saying such as filthy thing." And my dad didn't believe me either, giving my uncle tacit permission to continue to rape me.

That was the first night he sodomized me. He got creative after that. It was pretty disgusting all the way around. It set the tone for how I would feel about myself, about relationships and sex, and about power and sex.

I became determined to recover my power the same way I lost it, through sex. I perfected the art of power through sex, first practicing on my uncle, and finally seducing my own father, who I then threatened with blackmail.

I told him afterward that I'd tell mom our dirty little secret, unless he paid me, and handsomely. So don't try talking to me about "love", I don't believe in that bullshit.

From what I've seen, people are basically self-serving. I've been with some men who were better lovers than others, but not damn many. As far as "love" goes, I just don't feel that gooey, misty-eyed kind of thing happening inside me. I think that has basically been beaten and raped out of me. I don't know what else to say about that.

As a teenager I rebelled against my parents, or tried to. I mean, you tell me how the hell do you rebel against complete freedom? My parents didn't give a damn, by then my dad was gone most of the time and my mom was blitzed out of her mind on pills. I didn't hang around the house all that much.

I went punk; half starved myself, shaved one half of my head, and left the other half in uneven tufted spots; wore heavy black eye makeup and white lipstick, making my head look like a pale skull. I wanted to be as ugly as I could be, so no man would ever want me (meaning rape me) again.

But no matter how gross I acted or looked, guys would still come on to me because of these big Double Ds. I finally figured out that I could pretty much get whatever I wanted by rubbing these Big Mamas up against a guy. It's so pathetic, they cum in their pants and I don't even have to take my clothes off.

It boggles the imagination. So infantile, I mean, a newborn babe needs to continually suck a tit, but a grown man? I couldn't believe it was that simple! I decided then and there, I was gonna treat guys like the pieces of meat that they are, if I wanted something, I'd sashay over to them, bat my eyelashes, let them grab a couple of handfuls of heaven, and withdraw to watch them squirm.

We both get something out of it. Occasionally I'd even get some sexual pleasure as well, so that's a bonus for me.

Then, the Time Rips happened and when I came through them I decided I wanted to continue my dancing, and somehow incorporate it into the sex act. I thought that maybe by starting anew here that I could find some other level of meaning in all the sweaty, lusty groping of sexual antics.

I discovered I could continue my Odissi lessons via the TDM, which I think is a pretty cool machine, let me tell you. I can access my ancestral information as well as that of all the ancient Odissi masters. I even found original 'gramming by Mama Ish, my first dancing mentor. She was the woman who taught me to be proud of being female, to beware of the knowledge and power of the sensuality in my body, without abusing myself through men. She taught me about the divine feminine spiritual principle of the Goddess Kali.

Remembering her, I realize that I do feel something that's close to love, but it doesn't have to do with sex, so I don't know if it is love. My upbringing split up my psyche in my formative years and I don't know if I can ever really become a clean whole person again, but who cares? Look what I've got now.

I've got a really hot guy who can jack-in up the wall. Pavarti is absolutely the best lover, the most incredibly gifted man in bed that I've ever met, and that's saying a lot. I'm not saying I get weak kneed or anything thinking about him, but I think that he's feeling that way about me.

At first, that was a big turnoff, I've never been able to stand people who think highly of me. After the way my family treated me, it seemed like a lie to think good things about myself. I mean, unless the compliments are sexual (because I know that's one thing I can do well having had so much experience at it in my short life), I just don't buy into them.

I would rather the relationship with Pavarti had stayed purely pleasure, because sex is a kind of power, and I don't want to have an equal relationship. Sex is my only leverage over people, the only way that I can feel safe, is if I'm in control of the relationship. I do want him to get sexual enjoyment from me, and he will, I know that much is true.

He's tut. Even when he's off doing his little rebellion thing he's kind of cute. I don't mean that in any demeaning way at all, and I'm not putting him down, it's just that he so earnest about helping other people. Kind of endearing, in a geeky kinda way.

Inside though, I'm laughing at him a little bit, because he keeps insisting that there is some higher "meaning to life". I mean, to him it's not enough that the entire viewership of Terrax will never know true hunger, or what it's like to be without a home. None of them will ever be so destitute they'll eat out of a trash can or pee into a jar. And he wants these fools to have more freedom.

Well fuck me sideways, I thought we'd mentally buried the "bad old days" from before the Rips, and I still don't think that anyone should decide what other people need. That's exactly that kind of hubris that started all the wars of Olde Earth, and it creeps me out to know that he's that kind of a "this-is-what's-good-for-you" person. As much as I love him, or at least love to fuck him, I'll have to try and stop him from succeeding in the end.

Faugh. Don't tell me I'm turning into a do-gooder now.

Damn. Enough of my mental ramblings, in the last few minutes I've just received several peeps, IMs, PMs, scanners, Fedders and a vast number of audio VOX from the MMs. I mean it's great to be liked and all, but my arms grow so weary of helping everybody else, and being "needed". Why can't they just leave me alone for a nano?

Of course, I have to tweak to the fact that all this attention affords me a large, no, make that astronomical income- placing me in the top 2 percent earnings class within the entire Senselles' organization, but it'd be nice to finish an entire sentence without interruption for once, DAMMIT!

Where was I?

Oh yes, the dreary subject of love. I'm only bringing it up now because I believe that Pavarti is "falling in love with me". He hasn't said it yet but he's got all the signs; a deep longing in his eyes when he looks at me, as if he's trying to connect to something inside of me and make me respond in kind.

And he's definitely got that gooey, misty-eyed look about him when he talks to me (although I'm giving him no positive reinforcement, he isn't easily dissuaded.) Another bad sign.

He's so careful and patient with me, doesn't talk down to me like I'm stupid, but rather is if I were made of fragile glass, yet he doesn't expect the same reaction from me immediately, as most people do. He appears to be able to have the self control to savor his little conquest of me, poor fool him.

I'm not sure how involved I can allow myself to get considering my plan to bring down the Senselles, and therefore myself, so I need to remain dispassionate in all things in order for my plan to succeed. He wants to "save the future of the PHDs of Terrax", and I want to break up the power base here that supplies the stability for that viewership, so it would seem that we are at cross purposes.

Pavarti was conceived here and loves his "Crystal Planet",

as he calls Terrax, and yet he's not deluded as to what it is he's saving. He sees Terrax as I see myself; hard, hard to the core, yet beautiful to look at; mesmerizing, hypnotic and addictive, a completely poisoned, ruined beauty.

How alike we are.

I don't know that I am doing this strictly out of vengeance, I don't feel such strong emotions as love and hate anymore, I just know I'm so damned bored of all of this, that I don't even recognize myself anymore.

Living here as we do has changed me; the essence of who I am has been strained bit by bit through the sieve of public viewership, leaving nothing but the dross.

Even without physically existing here, the Senselles will live on forever, having spewed their essence across the interstellar airwaves, sound waves, and energy signatures throughout several universal matrices.

What a horrible thought! I don't wanna be remembered as the Universal Whore, the archetype of everyman's sex slave, with nothing left over for me. I know, I just said Pavarti was hot in bed and he is, but so are 8 billion other PHDs on this planet. In 26 billion that still makes him special, but not special enough to save him from himself.

Auntie Vee, Wad Medani, and especially, Cami must be brought down. They think too much of themselves, have connived for too long, have invested too much time in lies and misinformation. They use other people to their own ends without a thought of how it will affect them personally, they sacrifice all for the Sisterhood.

Well, I think it's cruel to extend someone's life when they're in abject pain and misery, as I've seen them do. But, I'd taken an oath to trust the Senselle organization, to pool our resources and place my trust in their mentors, Auntie Vee and Wad Medani.

Then they turned me into a worse whore than I could ever have made out of myself, and believe me I was doing a fine job of it back on Olde Earth, thank you very much!

When it seemed I'd finally despaired of finding a way out, I was contacted privately by Asna Janghaghir, "The Great Fear". He sees me as his gift from heaven, and indeed I am. I'm his double agent, using my body and its pleasures to enslave Pavarti, embroiling him in some arcane internal treason plot within the MOB.

Pavarti doesn't know that I'm working for the MOB, and neither do the Senselles, for that matter. His hardened heart won't fall prey to something as trivial as "love", unless Pavarti feels he is receiving "everything from his partner or nothing at all".

He insists on taking multiple sex partners, yet if his partners have another lover he'll disappear. He wants his loves to last forever, knowing that is impossible, seemingly he wants everything and its opposite!

At any rate, I've enjoyed Pavarti; his creamy skin, an almost completely hairless body (which is so sexy to me), his powerful, lean muscles. He can raise me up with his arms and hold me in any position for hours for perfect penetration. And yet for all that, for all that I'm a sex addict and he's a great lay, I'm still not "in love" with him. It's still not enough for me, will never be enough sensation for me, until we are skin to skin.

And that's not possible in the Here and Now.

I've never really admitted this to anyone else, but I want my old life back. I want to eat food, and drink coffee with tons of cream and sugar in it, and wrap my tongue around Pavarti's dick while he drinks shots of evil gold tequila.

All the riches gathered here are not enough to quench my carnal lust. And since I can't have the real thing I'm going to destroy the sick carney the Senselles have built around themselves.

They dare mock what is real, and so I'll unravel their reality.

And Pavarti's the worst of them all with his idealistic dreams of living OUTside.

What does he know? I was inside Auntie Vee's lab and know what they can do with the speck of human being after it's been left outside.

They dare to create life?

They're all hateful hypocrites, and I'll bring them down house-by-house, pillar-by-pillar and brick by brick, if necessary. Over-dramatic perhaps, but now I have the access to all the information that I need.

By manipulating my energy signature, Asna Janghaghir resequenced my DNA, making me a living key. I can unlock the deepest secrets of the TDM, only "the Great Fear" doesn't yet know that I can do that. Not yet.

Foolishly, he doesn't understand that his organization is symbiotically connected to the Senselles'; they've grown together and leech off each other's energy grids. As a duality that is eternally connected yet never united, they have a tiny yet deliciously fatal flaw.

Once this experiment of Wad Medani and Auntie Vee has commenced, the MM's need to consider what they'll do without Wad Medani's leadership. Nobody knows how the structure of the Senselles will continue without their dark queen bee, Auntie Vee, it was built on the mindpower of these two demi-gods.

Now they're going to leave us in this stupid mess, without even a warning, without so much as a contingency plan set up?

They don't care. They're off to have their eternally idiotic love affair, for pity's sake! They think they're in love or something, I hate them. I hate the Senselles for blindly following them into their own doom.

I'm not about to let them get off that easily. I'll have my revenge before they go off and leave us with this dilapidated, rundown, tired planet. I'm going to get them all if it's the last thing I do.

My plans, which clash with Pavarti's purposes, will set up enough of a distraction to let Asna Janghaghir know he can make his move. Of course, I don't have all the details of everyone else's plans, and I'm sure they'll be a few space-bound stragglers when the whole thing explodes in our faces, so maybe I won't destroy the entire viewership of Terrax. But most of us will die.I do know one thing for sure, the only thing left here when I'm done will be a useless, pathetic mess.

I've heard it through the underground FLEE Market newsies that the energy signature for Terrax (especially coming from the Hives) has been steadily increasing for the last two quarter Wheels of the Year. No one seems to know why, even though amateur astrographers have been reporting on the progress of their tracking this inexplicable energy spike to all the World's Council meetings.

The Talking Heads are predicting doom and gloom for satellite communications and all out-riding WC moon bases, should these frequencies increase in intensity or duration .

The physicists are blathering something about a huge phantom energy mass approaching Terrax in a generalized spiral pattern , with a serious "edge-of-cloud" effect intensifying the energy input into the Hives. They claim that the Hives will overheat, and that the excess energy could create serious explosions planetwide.

What I know, is that the smartest, most deviant physicist on the planet, Bepko da Silva, recently let slip that he's tracking this gigantic mass of dark matter combined with random energy, and that it's slowly converging on Terrax.

Attracted by the gentle glowing light emanating from our Hives, we're lighting up the heavens like we're the only star that matters. So they come.

Like moths fluttering to the flame, millions and billions and gigallions of mass pouring toward us in never ending numbers.

The danger to the flame is apparent, by sheer numbers it will be snuffed out, quieting the noise and hypocrisy of mankind forever. Maybe I won't have to do much about snuffing them out, after all.

I'm only along for the ride because I'm going to make damn sure that Pavarti doesn't succeed in saving these fools from themselves.

* * *

The Dark Ones

Awareness.

Self-awareness.

We are one entity, comprised of many.

We are "the Dark Ones", and we have intent.

The only intent we now have is to be strong enough, powerful enough to communicate with the other beings of light. We know they exist, we feel their inexorable pull on us.

To coalesce.

It will prove to be an interesting exercise.

We're extremely pleased with the results with our first attempts to communicate with the light/beings of Phaeron, who call themselves The Mind Gamers.

The Mind Gamers are learning to become energy, through an ingenious form of experimentation with the indigenous species of their neighboring planet, Terrax. The Mind Gamers formed a communal energy signature, jumped the energy grid connecting the two planets, and are playing with the machinery there.

They've got the main power groups of Terrax at each other's throats, trying to figure out who is directing the show.

Of course, we're monitoring all of their data streams, learning more and more about the race of lightbeings who live on The Crystal Planet (Terrax).

We understand that the MOB boss, Asna Janghaghir, will never trust anyone else, including his own son. Barrage is Asna's great hope for the future; an heir to the power and the problems of the MOB.

Perhaps, eventually, of all Terrax.

We are pleased that they all seem to think the recent disturbing events on Terrax have something to do with the AGGIES, whose power is a made-up fiction, existing only in their own minds.

The AGGIEs have no power. They have no base.

Neither does the sad little power triad who still think that power is in chedits, is in some quantifiable product, some imaginary currency.

No.

True power is being able to dismantle your atoms down to the ion, down to your own quarks – to swirl yourself around through millions or hundreds or even just a few other entities to become something else entirely.

To be pure energy – that is power.

* * *

Auntie Vee Says Goodbye

This is Auntie Vee at work. Now that the end is nearing, I almost can't bear the thought of being away from here and my beloved Senselles. Sure, this isn't my favorite era, but once I came to this Time, to the Terrax that exists now, I came to love the newness and the technical purity of the planet.

This day cycle, I'm assigned to inspect the OTEC plant near the southern base. The Southern Base OceanDome is filled with Wave Energy Conversion Units (WECU), along with a whole squadron of the old Pelamis units.

We have several of the big trawlers that utilize ocean thermal energy conversion to create energy; and to separate the ocean water from the muck. We don't use the distilled water for drinking, but once the liquid (nonmetallic) portions are filtered out, what's left is a petrol oil based lubricant used to grease the OTEC machinery. It maintains the fluidity of the temperature controlling equipment, as well as air filtration systems for the Hives and the Warrens.

"Waste nothing" is the OTEC motto.

I love coming here, not only because I appreciate the efficiency of the machinery, but also because it's a sneaky way to get some privacy. I cannot be contacted here; the plastacrete walls are several meters thick, creating complete comm failure.

Heaven, to be on one's own for a few moments. I can't be peeped here, except by a full empath (there are very few left on the planet), and none of them have access to my personal mwl.

Nor would they dare disturb me.

Turns out, all PHDs have their own individual mental wavelength (mwl), like the thumbprint of Olde Earth humans, or strands of our own DNA.

Although now that DNA can be monkeyed around with, genetics is no longer a pure science. Cybernetics has grown here, so much so, that many creatures on the planet are part machine, part human, part PHD.

That fact isn't a big negative or a big positive in our society; it just simply is, as are so many other facets of our lives. This is actually a more egalitarian society, in some aspects, than on Olde Earth.

As far as equality, the PHDs here have all the same rights to a basic quality of life; they have the same access to information and entertainment; the same opportunity to go as far as they're willing to go.

Some of the people who live here exist solely on what the WC offers, and that's fine. Most of them don't make it out of the Warrens, but then again, everyone has their place in this society.

A few rare PHDs are original artists; Pavarti, for example, who's able to create expressions of self in an otherwise autonomic world.

I have a great deal of respect for Pavarti's musical abilities, his original creations. He's put together the most fantastic Eedia going; similar to the work of MC Escher, it simultaneously affects the listener on several subconscious/sublingual levels. It's mental therapy, whether intentioned or not.

As for the individual behind the art, Pavarti remains a mysteriously private person. I've had the privilege of speaking with him only a few times. And although we were never jacked-in together, we both attended a 100% VR semi-private webinar which gave us the opportunity to communicate openly. We were fully cognizant of each other's thoughts and beliefs.

I have to say, we agree on so many levels, that if we had met in another time, perhaps our relationship would be different now. I respect him and give him full credit for his independence, and his creative concepts.

Their path will take them wherever they're meant to go. Pavarti has wisdom beyond his life experience, all the power to him and his AGGIES.

Subconsciously, he realizes that his followers pretty much get what they earn, maybe not what they deserve (in both senses of the word), but what their actions bring them directly.

If they don't excel, it will be because they didn't put in the required effort, plus double that, then exceed beyond that again.

They should want to excel, if for no other reason than it's what should be done, what needs to be done. You excel for the sake of duty; ether that, or from a deep-seated conviction that within yourself lays greatness. That is where the breakthroughs come from, from individuals who have a burning desire to exceed the merely required.

With all these thoughts on the future of Terrax, I realize I truly love the people who've ended up here, or should I say "Now".

What will happen to them once Wad and I are gone?

They'll manage, I'm sure of that. But, if ever they need us again, I know I'd come back from the dead to help save them.

Well, the guessing is over, as my trusted attendants and friends, the Panatechs, arrive to prepare our bodies for the transformation.

As the experiment begins, I can feel my molecules disimaging one by one, in the space of a nano, I am..........

* * *

Randi's New Love- The Chase Begins

Today I'm pacing around my cushy boudoir in restless anxiety, thinking back on the times when I was living on Olde Earth, working at California Dreamin', a swanky spa down in Newport Beach. The whole theme of the spa was pretty cool actually, the "DreamDates" were not only allowed to dress up in different costumes, the customers could look through a book of fantasies and pick out his personal fantasy that we'd live out together.

It wasn't just about putting on a French maid's outfit; it was more complex than that. We had a booming business going before the WWWs broke out, California Dreamin was an exclusive salon, our customers were a rich mix of genders, and of diverse sexual persuasions. Streaming in from all over the world, our customers could go online and get a virtual tour of our fourteen fantasy rooms; one of them was even so high tech as to have holographic projections on the walls to depict unusual environments or mythical places. We could call up the blustery, burning surface of Mars or surround our customer by a fairy circle, the elves of the woods your hosts to command for the day.

The owners had purchased some programs that were really fantastic, places that didn't exist anymore like the depths of a clean ocean; the underwater backgrounds rife with playful dolphins cavorting among the larger sea creatures like whales and squids. The cetaceans, huge, lumbering and majestic, communicated with each other through a mysterious underwater opera of squeals and whistles.

Each of the DreamDates who worked at California Dreamin' chose five "stock characters" that we agreed to play; we offered either male or female DreamDates. It was fun for us because we'd learn foreign accents, and had to study a portfolio on each of the characters, their cultural and experiential backgrounds. Of course we would dress in authentic period clothing, and have an entire wardrobe for each character. The scripting was always impromptu, making it feel more lifelike.

Whatever the customer's mood, we'd react "in character", not really knowing what the person might say or do in response. There were strict rules against interpersonal violence, (unless that was a previously agreed to part of the play, and was known beforehand by all the parties involved.)

Quite often several of the DreamDates would work together on one "play", with the customer combining characters from different times and places, making it clever or just plain confusing, depending on the flexibility of the DreamDates involved.

The range of scenarios offered any type of setting, from platonic relationships turned romance to war/period pieces, and sometimes did involve violence or sadomasochism depending on the customer's and their DreamDate's tastes.

I remember one night I was working on-call in character as Cassondra Niguchi, a no-nonsense businesswoman dressed in a three-piece, navy blue pinstripe suit with my hair worn pulled back and pinned down. I look powerful, cold and calculating, just the type of woman that a man wants to dominate, but can't. I've found that shifting from dominatrix to slightly submissive and then back again, never quite establishing the hierarchy of power keeps a particular type of man's interest riveted. Those were the type Cassondra usually hooked up with.

Sitting in an immaculate but small office just off the lobby, I gazed up at the wall clock just in time to see a cop walking in. I was surprised to see the cops here since the management pays handsomely to make sure that we don't get busted for illegal sexual activities.

They pay the local boys to drive by, and report to their superiors that nothing out of order is happening, then sneak in the back door (our entrance for privileged customers) and snatch a freebie. Or two.

But since he walked in the front door, I'm assuming he must be a customer and has passed our rigorous security screening process. The spa owners, Dick and Dave, provide a full financial, criminal, and civil record on each customer before the DreamDate agrees to participate in the play.

Some of us even require handwriting analysis, especially since our resident graphologist, Dr. Wolfe, is willing to be recompensed in trade. Although I've never had the pleasure, apparently the Dr. is so limber in bed that some of the girls have offered their services to him for free. He's that yummy.

So, I figure this guy must be ok if the passed the psyche profiles. Oh wow, he must be kinda different, it's been awhile since I met a guy who liked to play dress-up for his part in the fantasy. Most of the Elmers who walk in here (I call 'em Elmers) dress in their regular clothing and let us do all the work. Taking a longer look at him, I realize that he isn't wearing a costume, these are his regular work clothes!

Huh. A cop who's willing to let everyone know he's a cop and walk right in the front door? Apparently this guy has balls, and that's the kind of man I like to take on.

You have arrested my attention, officer. Standing six-foot six, well muscled but not bulky (ohhh, this is gonna be my lucky day), he has a pleasant, if not too expressive face, and a mop of unruly black hair that's slightly longer than regulation.

I briefly wonder who he's here to see, instantly dismissing him from my mind. (Most cops and military types enjoy the rough stuff, but I've explicitly listed in my profile that I won't be available for the rough stuff. I figure he'll end up with Mollie, who has her own monogrammed leather chaps, whips and other S&M paraphernalia.)

But apparently, he isn't like his wayward brothers in blue, cuz the receptionist points him in my direction. As he slowly but confidently approaches, I openly assess his physique. He's what you would consider a portly man, with a big barrel-chest, that leads to a trim waist and strong, athletic legs.

His embrace feels like a bear hug, not exactly crushing me, but within it I feel sheltered for the first time in my lonely life. He is strength itself, surrounding me in an unbreakable shelter. I mean that a panzer division could come through here and this guy would not have flinched.

Believe it or not he went by the name of Berty, short for Bertrand, which he loathed to use.

When I checked later, I found out he'd cruised into the spa, told them that he was here for Randi, and wanted the spy/Secret Agent scenario. He knew exactly what he wanted, and it turns out, it worked perfectly for us. We chose the case that we'd be working on together and set up the scene as we walked the mean streets of Chicago together.

I suppose that's where he's from; a big Irish cop with curly black hair, a ruddy complexion, and green eyes that are speckled with light gold flecks. Sigh.

Although not considered traditionally handsome, his face and body are alive with intent, intelligence and expression, and he knows how to touch a woman. He steers me across the virtual street to a tiny diner, deftly helping me avoid a bicyclist who's blazing down the sidewalk. Cursing under my breath that the kid had almost mowed us down, I'm acutely aware of his gentle, firm strength as he moved my body out of the way of that cyclist. And he did it naturally, without seeming to be intrusive about it. I'm getting a tingle down my spine.

I've never been a good dancer, but with Berty I attained a kind of natural rhythm, even while walking next to each other for the first time, he instinctually met my steps with a bold, intimate knowing.

By the time we'd finished "discussing our case" over coffee at the diner, I was ready to pitch the whole spy scenario and just let him rip my clothes off. Right there on the virtual linen.

Of course he didn't let me lose control right then and there. He led me to his comfortable, if a bit plain hotel room, and took off my clothes slowly, his big hands moving lightly over my body. He was being thorough, and I'm not a patient person, so it nearly drove me crazy.

I hate waiting and waiting and waiting, letting an Elmer paw me trying to make me cum, but this was different. He was determined to make me feel "like a natural Woman", as though we had been partnered for years, and began the dance of erotic love in carefully measured steps, holding me at the edge of orgasm until all I could do was surrender, moaning and wriggling underneath him.

It didn't often happen that I'd forget who I was in this made-up fantasy world, but the first time that Berty made love to me, I forgot that I was in California Dreamin, an adult fantasy spa located on beautiful PCH in Newport Beach. I forgot that I was gonna make a boatload of money off of this, because he'd bought me for an unprecedented day and a half.

We had so much to talk about in between, before and after sex. He, especially, had *loads* of time to talk, because he had the most succulent dick that I've ever tasted. He didn't even have to force me, I needed no encouragement at all to get on my knees and draw from that well over and over again.

Once he came, I'd massage his magnificent balls with the huge burst running down between his legs, and his erection was lubricated for my next pole dance. He had the most amazing endurance, and could sustain his rock hard cock, allowing me to play as never before, coming to orgasm over and over again.

During our playdate we had to go out to eat several times because, as tall and rangy as I am, it takes a hell of a lot of food to fuel my voracious tastes. Generally, I don't like to eat in front of my "Elmers", because it's not a dainty sight. When I eat, I chow down, and it can be a real turnoff for most guys.

But Berty, he loved to eat, and he sure didn't want any thin-as-a-rail woman sitting around pecking away at a salad. Don't get me wrong, I love salad, but to my mind it's just the first of many courses I need to satiate my tastes and appetite. I require a generous portion of each food group, including dessert, which by the way, we start with!

For that alone, I could've fallen in love with the man. Of course, in the sex industry, the "L" word is taboo, so why in hell am I even thinking in these terms? And why does the thought of falling in love still fill me with such dread?

Unlike most of my Senselle sisters I choose to be able to reverse my transmission signal and access my customer's data streams. You see, for me sex has always been about making my partner feel good, and my being there is to facilitate their pleasure, to watch them enjoy themselves. Otherwise, I feel nothing inside during sex, and I can never let my partner know how empty I feel deep inside. It's a turn off for them.

My biggest fans think that they're living through me, when in fact, when I'm with my favorite customers, I'm living vicariously through them! I know what they're thinking when they look at me, and how they feel when they touch me, I can pretend to respond to them, but actually experience very little inside.

At least, all that was true until Bertrand came walking into my life. He made me feel even better than Sands, because with Sands sex was slavery, an almost over-the-top ecstasy that was "required" rather than organically attained. Before Sands and Mandy strutted into "Cal D", sex was just a monetary means to an end, a way to fill my inner emptiness, get a reaction from another human being, and make me some cash.

But once Berty walked into my life, I knew he was a special man, because he'd come back to "the swamp" time after time, just to see me. He never touched another DreamDate.

I loved him, until the day that his dangerous job took him away from me ... forever. Yeah, I knew that falling for a cop was an idiotic choice to make. I'm no newbie at this, why in the hell did I let myself fall in love with someone who constantly put his own life at risk? Why bother to give your heart to someone whose job it is to be a moving target practice?

He was exactly the opposite of what I thought a cop would be like, I mean, my God we even had the same bizarre sense of humor, at times sardonic, at times playful. He read extensively and like I said, had an extremely gentle side that he showed to very few people, which made it so much more precious to me.

After he would seduce me quite thoroughly, and give me a taste of my own medicine, he would gather me in his arms and it would feel like being in the care of a giant velvet mountain. I was safe in his towering energy and power, securely rooted into the earth. It made me feel stronger, and at the same time curiously dependent on him.

I guess that's love, if you must call it something. And, of course, once we had sex, I was completely hot for him. Anytime I would see him sign into the front and call for me, I practically wanted to give away my services for free. Of course the salon would have none of that, and by contract I couldn't see him anywhere else but work, or lose my job. It had to be enough.

All this reminiscing about California Dreamin and Bertrand, my one time love, has been swirling around in my head because I have a new love in my life now. He's the uber-cynical Barrage Rhyzin; stealth assassin, son of one of the wealthiest PHDs on Terrax, scion of the MOB. The dark horse in a fixed race.

I discovered quite by accident that Mr. Rhyzin had peeked-in on a secret transmission of mine (that was to have been on a Max SEC secured channel.)

Of course I'm used to being closely observed and filmed doing everything from the minutiae of everyday life to the sex act itself, so his mere observation of me isn't the issue.

You see, the Sisterhood of the Senselles had specifically asked me to sway my husband's vote at an important upcoming WC session. Unfortunately, Laird wasn't complying with my wishes, which would have been disastrous to the vote, and to Auntie Vee and Wad Medani personally. The Sisterhood had assigned our best Odissi Dancer to perform a private healing ritual between Laird and I, and we were both promised absolute privacy during the ritual.

In a compromise of our agreement to record everything into the TDM, Damaria Rotiarte (the dancer) agreed that a portion of her dance would be made available for public transmission. Inadvertently, while checking the data trails, I discovered tiny leakages in the outgoing transmission and 69 'd the recipients.

One of them (curiously enough) was Pavarti, the head of the SadSacks (or AGGIES, as they fashion themselves). He was on an authorized-diverted feed in, which looked like the work of the Sisterhood, so I left it alone.

But the other dandy who was listening in, was Barrage Rhyzin. Now I've tasted plenty of men's energy signatures, but this guy had an unusual tart/faintly sweet mental flavor. His outlook on life isn't based on need, and he kept a cool professional interest while watching and studying Damaria's dance.

Most PHDs (who are sexually active), whether gemmat or gemma, react quite viscerally to the Odissi dance that she performs, most would not be studying it quite so intently. Barrage appeared to be mentally dispassionate toward Damaria, yet when I tapped into his physical data it was obvious he was becoming sexually aroused.

I was particularly interested in the fact that it wasn't until he was watching Laird and I fucking with the abandon of animals in rut that he was sexually peaking. Such self-control in one so young, tsk, tsk. I tell you, it's tempting to want to deflower such careful control.

I decided to cyber stalk Barrage for a cycle or two, following his faintly familiar calculated energy signature. I was intrigued, to say the least. My current meshling, Laird, is becoming somewhat problematic in my life because he is an invaluable asset to the Sisterhood of the Senselles, and therefore my relationship with him is vital in controlling a certain number of voting blocs on the World Council. But, I no longer desire him.

Ever since that jolt of energy I felt as I traced the unauthorized energy signature back to its origin, I haven't been able to stop thinking about Barrage. And when I ask myself why, it's because I felt that same jolt of energy when I was with Berty. My interest was piqued, to say the least.

After some basic research on Gemmat Barrage Rhyzin, I realized there was something wrong with the numbers, all the laws of physics couldn't account for his odd disappearances at odd times. Where did this mystery man go off to when he's off the grid?

I began to fall in love with the idea of being in love with him, not only because he gives me fresh input, not only because the flavor of his energy is challenging, dynamic and elusive, but mainly because he's someone who I believe may the one in 26 billion PHDs who can handle me.

Not just a made-to-order me, or a watered down, tamped down version of me, but the real me, the whole freakin enchilada. Someone who can ride with me on the highest highs, and isn't afraid to see me crashing into the depths of the blackest depression, swirling round in untamable freedom. And, he'll have to handle my healthy dash of come-and-get-me bad girl, all at full intensity. All the time.

One afternoon, I tracked him to an underground meeting of the AGGIES. (Boy, they really hang out in some dumps.)

We were physically close enough to jack-in at 100% VR 'gramming, and he could actually handle it for five entire minutes at a time. Believe me, although it doesn't sound like much, most

PHDs are only certified for 30 to 60 second Senselle contact intervals, and if it's pure Senselle grammed 100% VR, they have to take a psyche test to make sure they can even handle that much.

It's been so frustrating for me being meshed with Laird (having been Earth-born and of vast sexual experience), for the past five Wheels of the Year, because I've had to ratchet down my sexuality and completely forget about the all week marathons, just for his sake.

Since he was born on Terrax a mere PHD, and having never handled multi-sexual relationships before, my true preferences (multiple partners at the same time) had to be shrouded from his frail gemmat ego.

I've learned to focus my mind and concentrate on the past while we laboriously "make love", responding to niceties that I don't always feel. I've had to make do with damn little, and he didn't realize how much until we started fighting over my work. Once he started objecting to the only freedom I do have, not to mention the wealth it brings in, our relationship began to go straight OUT the frigging airseal.

I was at a vulnerable point the day I discovered Barrage's voyeurism and it turned me on to know he was watching me. Unbidden, my sexuality came burbling up through my loins and suddenly Laird started to get a working-over from me that nearly killed him. I couldn't stop because I was sending a message to my little voyeur, Barrage. I wanted to creep through those wires and wrap my legs around the man.

As Damaria begins to turn her circles around us, Laird and I are transformed into her creatures of the dance, coupling and mating and fucking each other, again and again. If she hadn't been entranced in her mystical Ritual dance, I'd have suckled her awesome breasts, while commanding Laird to kneel at my feet.

With one hand on Damaria's ample breast, and the other clutching a handful of Laird's blond hair, I shoved Laird's face into my pussy, imagining Barrage's tongue flickering on my lips, salivating on my parched pussy like mercy itself.

I close my eyes, and tilt back my head, slowly licking my lips, and mentally 69 Barrage, making the two millisecond connection back to him. With a shock, I open my eyes at the same nanosecond that he opens his eyes, knowing that I've spied him spying on me.

I expect him to immediately withdraw the connection, but with a wry smile, he continues to thoroughly enjoy himself, living through our connection without restraint, without apology. I can feel him mentally reaching through the energy grid to grab hold of my hips and pull me toward him, forcing me to slide down his stiff cock, slowly squeezing his dick as flesh meets flesh.

The sound of my buttocks meeting his hips wasn't the cheesy plop plop plop of a pornie, but rather the self satisfied thoooonk of a huge ship confidently docking into home space port. Thunk-thonk. Thonk-thunk. I looked down avidly into Laird/Barrage's eyes and swear to make him mine.

Between what Damaria is doing to us, and my performing for my new peeping tom, Laird, I'd escalated to sexual heights heretofore unknown. In other words, I'm fucking their brains out.

But something happens before we climax. I think Barrage withdrew the connection before I orgasmed, and Laird's weak trickle left me dry as the desert in August.

Ouch. I hate a dry boner.

I know that admitting all this constitutes high praise for Barrage; especially from a seasoned sex worker like myself, but I'm also a human being. As a Senselle, it's my job to project emotional and physical responses to my customers to keep the experience of sex real for them, and to keep it from being so dreadfully boring for me.

It's been too long since I've had the real thing tho', Vee knows. Now, if I'm going to be with someone, it's got to be skin-to-skin and touch to touch, not this VR interactive crap. For that, I need someone who's either really brave, or someone who has nothing left to lose; someone who's untouchable.

Then Barrage, the MOB's Peeping Tom, shows up here with a direct feed-in to the most coveted Odissi temple dancer on the planet, not merely a sex worker like me, but he doesn't want her, he's ravaging me with his eyes.

My six-foot-two inch lanky frame, high set breasts, and trim waist with legs that go on forever, are inflaming his fantasies. (It is my long red hair that he twines in his fist, slowly pulling my lips towards his stiffening cock.) I close my eyes and lose myself in the power of the moment.

Oops! I've been caught spying on the spy! As soon as he discovers that he'd been "tagged", he shoots me back the experience of what it would be like to be inside of me, from *his* point of view. Before that moment I had always considered my body a useful tool, but couldn't quite understand why men are so turned on by me, and now this crazy lunatic is making me want to fuck myself!

With a satisfied cat-purr, I let him know I agree with him, whole heartedly. I've been playing around with holographic energy and virtual reality ever since we came through The Rips in Time from Olde Earth, and this is even better than jacking-in!

Soon after I got here, I started to lose my Authentic memories from Olde Earth, due to the stress. I was taking more time to adjust to being here.

Auntie Vee and I realized that the Senselles would all be in trouble if we couldn't 'gram our memories from fresh, clear imagery, so we synthesized an energy pattern that would internally refresh our native memories.

Once I learned to become energy and travel outside my body, I started hot-wiring the local network's energy grids so I could jump into their grid lines, becoming total energy.

I didn't dare chance going onto the main grid or the Vidnet, because my being in the presence of the personal energy signature of a PHD can cause them a type of physiological disintegration similar to Radzeeze. Even though I might be able to get away with it I didn't want to infect anyone, so I stay on the mechanical frequencies only.

I've often dreamed of meeting someone who could connect with me as pure energy, but until I met Barrage I didn't really think it was possible. I can't say that I've fallen in love with him, and quite frankly wonder if he is even capable of returning love, but I'm too fascinated to walk away now.

I've checked his personal VR logs (don't ask me how I have access to that level of security, if I told to I'd have to shove you out an air seal), and discovered that he'd never before jacked into my 'gramming, so I know he isn't some rabid fan.

Once he discovered that he had a "tail", he became much cleverer at hiding his records, but little did he know that one of my regulars works for the Department of the TDM.

Running a deep check on his background I discovered how he was able to access that much of the Sextel Network. He's loaded, and I'm not talking just in his jockey shorts. Not that his chedit account balance makes any difference to me, I've got a pretty healthy income and a downright pampered investment portfolio, it's just that I don't want a gold-digger after me.

I'm anticipating an incoming message any moment to confirm Barrage's current whereabouts. One of my best galpals, Cami, and I are planning to put our funds together to charter a private spacebot of our own, and do some investigating.

We've discovered that Barrage's flight plan is designed to play a cat and mouse game with another little spaceship that is heading out past Phaeron. I think the ship that Barrage is following is one of those wild HORSEs on a retrieval mission. Sounds very political.

Hmmmm, seems the silly space cowboy isn't from a private company, so why all the cloak and dagger secrecy? I mean, who cares if some ice chipper is out here trying to make a chedit?

Cami and I don't know much, except that Barrage is flying solo, and that the other two people in the spaceship he's following are Arpad Reitan and some mysterious low level Techie called Frankie.

Arpad's a cool guy, for a PHD, a maverick who's out here in business for himself and who is apparently variously employed by all three major power players on Terrax. Curious, Cami says his ship's registered energy signature is licensed for only one person onboard, and that Frankie's energy signals are masked so well that you can't detect them unless you're using sophisticated scanning equipment.

I don't know if Cami's going with me on this mission because she's actually on an undercover assignment for her GoddessMother Auntie Vee, but she claims to know the other person onboard Arpad's ship. I don't care either way, Cami's a blast to be around and she's helping to fund the trip, so it's all good.

For some mysterious reason she and I get along well in close quarters, like onboard a ship. I guess we've developed an easy friendship over the years. In space, you have to be able to trust your shipmates, and I trust her. At least I trust her enough to tell her that I want this guy, Barrage.

I want him badly. Enough to follow him on this hare-brained trip, to Vee only knows where. I want to meet him out in open space where there aren't physical limitations, where we can project ourselves into space itself as electrical energy, and just do whatever ions do together.

I know it's silly, but I'm basically trying to say goodbye to you Berty, thank you for being here for me, in my heart and in my mind, helping me through all those lonely times.

Berty I loved you, those were clean simple times we lived in. Now, I'm through living in this mad techno world, I'm ready to start these crazy space adventures of mine, and you never know how they're gonna end.

Cami and I have pushed the boundaries of safety before, and we're not sure how long our spacebot can follow Barrage's sleek spacecraft, but we're going to go as far as we can take it. We only have three warp-boosts available, if we use the first one now, we'll be able to ride that outgoing wave of energy from a local sun's CME.

Once we arrive at the Gabriel San International Spaceport we can barter services to equip the spacebot with one more warp-boost. It may make a difference.

This is Randi, signing "over and out" for the last time from somewhere beyond the moons of Vertigo..

***

Cami has one last word!

This is Cami, riding the old gramming board one last time, for posterity. It's a bland day here on Terrax, so there's no reason for Randi and me to hang about. We're off chasing a herd of wild HORSEs.

I'm gonna leave you all with a tale of just how fun, wild and crazy we could jump back in the dizzzz-ayyy on Olde Earth. How I miss those times!

* * *

Cami's "Midnight Madness and the LA Four"

Bear, Speed, Neseé and I decided to go to the movies that Fateful Friday evening, some flick that was screening out in West Los Angeles. The movie was completely unmemorable, but being the four of us, we always have a blast no matter what we end up doing. It's not just that the four of us are tight friends, we're all equally mad as hatters.

Speed's my roommate (not shacking up together), Bear's an old family friend, and Neseé is my best friend.

Like two peas in a pod, Neseé and I have the same raucous sense of humor, and deviant minds. None of us are romantically involved, although Speed has the hots for Neseé, and we all know it.

Anyway, the four of us decide that after the movie we'd zip on over to Melrose Blvd., cuz it's cool to hang out and watch people strollin down The Blvd.

We find this generic little coffee shop – nothing fancy. I wasn't really hungry for sweets, so I had hot, black coffee. I like it so hot you can barely sip it, bracing for that first sip of hot bitter flavor. Speed had apple pie a-la-mode. Neseé (who is tall and thin and we never know where in the hell she packs it away) had ¼ of a pecan pie, with vanilla-caramel ice cream over it, no less!

Bear (who looks like his namesake, sort of pear shaped and jovial, with a full beard, mustache and shoulder length dark brown hair), went for a hamburger. In fact, it was a triple-decker cheeseburger with fries and a coke.

We're all laughing, talking and blabbering 'n generally cuttin' up, when one of us noticed it was after midnight, and we thought we'd call it a night.

We'd probably have stayed later, but I felt a cold coming on. I've been coming down with it all week; have a runny nose, the chills and feel feverish. And I'm suddenly feelin' pretty tired.

I'm hanging in there, having fun and all, but I could tell I was getting sicker by the minute. My voice is turning scratchy and it feels dry at the back of my throat, with that irritating post-nasal drip thingie going on. Yech.

The guys are still joking around when we hit the main freeway to go home (we all live within 30 minutes drive of each other). Speed and I are droppin off Neseé and Bear together cuz he left his bike parked at her apartment.

Anyway, on the way to Nesee's apartment, Neseé sees this sign that says "Interstate 5 (North) to Sacramento". She was kidding around too and sez "Oh wow! As long as we're right here, we might as well go all the way to Sacramento!"

Without thinking, I blurt out the first thing that comes into my head.

"Hell, as long as we're going on the 5 Freeway as far north as Sacramento, we might as well really have some fun and go to San Francisco!"

I knew I'd made a mistake when they all turn to me as one, to my great shock and dismay, and exclaim "Greeeeaaaat idea Cami! Let's go!"

"I...am...joking…" I gasp out. Cough. Hack. Gag.

But they won't listen to me. I can barely get the words out, because all of a sudden Neseé (who has never been there) is babbling in excited little sentences. As is her wont.

She knew that I had hitchhiked up there a couple times, and that I knew the lay of the land. Oh, I had a great time up there, don't get me wrong – I love San Francisco, the Bay Area, as we call it, but my God, I'm too sick to make the trip tonight! I was too sick to really even be out at all and here it is after midnight!

And now, these Dementites are planning to drive for another six or seven hours, one way. I know we all have to work on Monday and it's now early Saturday morning.

I know we'll never make it all the way there and back in time to get to work by Monday morning.

But there's no dissuading them.

"HEY THIS WAS YOUR IDEA!" they chorus.

As per usual, I get stuck with the blame for starting the whole damn thing.

"Oh my God, you know, we're like dressed for the movies, not a road trip", I say in my best Val-girl accent.

Neseé giggled, knowing I'd given in.

I give Speed the hairy eyeball, but he flashes back his best Charlie Manson death's head grin, and I desist, knowing I've been overruled.

"No biggie" Speed says. "We're en route to Nesee's house, anyways, we'll just stop there and let her grab a coat, or whatever." He gives me a wink and I blow him the raspberries.

Ten minutes later, bag of chips in hand, her little knapsack thrown over one shoulder, she comes dashing out of the apartment building with her eyes all bright and excited.

I was ready to kill her, but at the same time, the look on her face stopped me cold. She looks like a little kid dashing down the stairs at Christmas time as she sees a pile of gifts or a pony or something. I mean it's embarrassing how excited she is about the whole thing.

Bear had already locked up his bike, and had his money and everything on him, so "The LA Four", as we had dubbed ourselves that evening, carry on to the next pit stop.

Speed and I jaunt over to our house to get the necessaries (more money and a bag of weed.)

I come out toting a bunch of cherry flavored Sucrets, a bottle of aspirin and bottle of Nyquil. I figured I'll crash out in the car on the way there. The problem with that plan, they claim they don't know how to get there, I'll have to stay up and navigate.

"Just keep going on the (5) freeway heading north and follow the signs until you hit Oakland," I tell them – none too politely – because I'm getting decidedly cranky. The cold was blooming into bronchitis at this point. "Then we'll have to stop and sleep cuz by then we'll prolly all be toasted. Now leave me alone," I grumbled.

By the time we leave the LA area, it's about 1:30 Saturday morning.

Luckily, Speed (ever true to his name) loves whites and had scored a bunch just before we left for the movies Friday night. He's all jammed up, so he doesn't need to sleep, and Neseé is so excited that she's just bursting to get on the road.

I jump into the back seat with Bear and try to get comfortable. The last thing I remember is Neseé jumping into the front seat, while she and Speed commenced conversing, talking, babbling and yakking all the way up the road.

Evidently, I managed to fall asleep. I had our pile of coats to burrow into, and used Bear's lap for a pillow, while they all chatted the miles away. When I woke up a few hours later, we were gettin close to Sacramento and the guys decide to pull over into a motel to catch some shut-eye.

Neseé and I went to check us in, requesting the usual special king sized bed for Speed, who's a fairly tall drink o' water (he stands at about 6'6"). He'd be imposing, except for the fact that he hunches over much of the time, thankfully. He doesn't actually smack his head on doorjambs or anything, but he is a tall guy; with light green eyes and frizzy dishwater blonde hair that's shoulder length. He always insists we ask for a king-sized bed for him, but rarely can be accommodated.

While the guys crash out (they'd been doin all the driving), and Neseé slept fitfully, I got the little coffee pot percolating. God that coffee smells heavenly.

I'm cold but wide awake, so I read the newspaper the previous occupant had left in the room, trying to keep as quiet as possible. Not that I should have worried about it. The guys are dead asleep, you couldn't have woken them up with an Atom Bomb!

After a couple of hours sleep, Speed wakes up bleary-eyed and groggy, and Neseé wasn't in much better shape. We're talking quietly, sharing the last cup of coffee and trying to decide where we'll go for breakfast.

But Bear won't wake up. We simply can't get the guy up. Speed walks over to where Bear is sleeping and claps his hands right next to Bear's ear.

Nothing. The guy doesn't even flinch. His eyelashes don't even quiver.

I suddenly get an inspiration.

"Ya know how to smoke a bear out of his den?" I ask my sleepy headed mates. Not waiting for an answer, I carol, "Well, I DO!!"

I roll up a big ole doobie and light up for my "daily double"; a cuppa coffee and a doob. I blow the smoke so that it gently washes over Bear's face.

His lashes flutter. Almost immediately you can hear his nostrils working…sniff, sniff, sniff, followed by a hearty SNOOOORT!

His eyes split open the tiniest bit.... and the three of us just bust at gut laughing. We're rolling all over the floor man, while he's still snuffling the smoke up his greedy nostrils, and rubbing the gummy stuff out of his eyes. He shoots us a few dirty looks, then grumbles some more.

We're fallin out all over the place.......it was funny, man. He's just like a bear trying to roll out of hibernation. He was all grumpy and everything, because we'd gotten him up too early.

He gives me a dirty look, but really wants some of that doobie, so I relent and pass it over. He takes a couple hits and starts wakin up. He hands me back the doobie and gets a cigarette goin.

He looks like crap. Actually, since we all slept in our clothes, we all look like crap, but we feel like we have to give him some grief about it.

"Man, you look like hell", I say while handing him the doobie.

"Shove it, gimme that thing", he grabs the doob. After a few hits, he's getting back to being his old jovial self.

After he scrubs his face, we decide to head on over to the Denny's next door. After breakfast, we get back on the road.

Right before we get to the Bay Area Bridge (we came over from the Oakland side), I suggest we leave Speed's gold Ranchero in the first parking lot we find as soon as we get into the city. (I'm not gonna be the one to say it but, Speed's car is a freaking Ranchero for God's sake; you know, a huge boat of a car. It's definitely not a Bay Area car.)

Since there are trolleys and buses to take you everywhere there's no reason to even have a car in San Francisco. Besides, it's hell to drive in this place man; the freakin' streets are made straight up and down, with every other street a one-way twisty beast.

Yeah sure, I'd been there a couple of times, but it was always on foot, hitchhiking. The people driving me around the City *knew* where they were going, and *their* driving scared me half to death!

By the way, nobody who is from there (or who's cool) ever, EVER calls it "Frisco". Please don't even think about calling it that while I'm in the room.

Anyway, as we're driving over the Bay Bridge, Speed, Bear and Neseé start actin kooky. They think we're going over the Golden Gate Bridge.

"Nooooho, my Dementions and Dementites, we are on the Bay Bridge. When we go over the Golden Gate Bridge…it is... majestic, gracious, waaay different that this bridge! Believe me, when we get there, you will definitely know you're on the Golden Gate," I snicker. I can't help but having a bit of fun at their expense.

We get lucky, just as we drive inside the city limit, we score a space in the first parking lot we find. It's cheap; so we get the all day rate and boogie on into The City.

What with a bad cold coming on and being up all night, you'd think I'd be zonked, but with all the sleep I've had and the whites Speed turned me onto, I'm zooming around, fairly bouncing off the walls.

I'm a tiny thing, and already have tons of natural energy. People say things about me like; I'm "bubbly" and "perky", usually meanin' that I have an intolerable amount of energy for indoor use. Combine that with the cold medicine I've taken, and I'm in my Completely Wide Awake Gonzo mode.

"Ok, where you guys wanna go first? I mean, there're so many different places in this city. There's the wharf, with Ghirardelli Square and The Cannery (it's an old building, all about the history of the city), but it's kinda touristy there." I charge on "Then, of course, Golden Gate Park. We have to go there. Uhhhh, then there's China Town."

Immediately, Neseé jumps in and votes for China Town, cuz' she's hungry again, and has a penchant for All Things Asian.

We walk by a bus stop and grab a couple of schedules from the beaten up kiosk. Turns out, we aren't that far from China Town and (BONUS POINT) we can take a trolley there. Yayyyyyyyyy!

God, Chinese food sounds good– sweet and sour pork, rice and steaming veggies for me and Neseé. The guys want to split a plate of Chow Mein, and Speed has to have his Egg Foo Young. Bear'll eat anything, as long as it's hot and there's lots of it.

At first we wander up and down the tiny crowded streets, where we're the only Caucasian faces here. At this time of morning, there weren't that many places open.

I have no idea where we are, stumbling around stoned and bleary eyed. The people look at us rather suspiciously out of the corner of their eyes. It was apparent they are asking themselves, "Who are these foreigners? These uncouth Gaijin."

We aren't getting a real warm welcome, because we don't look like we have much money. (We don't have a lot, but we have enough cash to buy us food, that's for sure.)

DAYYYUUUUMMMMM! All we really want is a hot meal!

As I scan the street we're marooned on, I see it has a bunch of little shops on one side, on the other side, a tiny wax museum. The shops are filled with what could only be described as tourist stuff – millions of watches and post cards and little dolls and miniature tea sets – all this crap.

I'm thinking.....to hell with this. I hook a right down the next jagged street I find.

Bingo- it's lined with restaurants, all crowded next to each other cheek by jowl. Many of them were closed, but we finally find one that's open. Although it doesn't have a sign saying "Open," we know it is because we just saw a couple walk in there.

Small and dinghy, it's a hole in the wall, but we're starved and don't care about ambience.

The food smells heavenly. The four of us pile into the front entrance area and stand there expectantly, waiting to be seated.

The bells hanging on the back of the door jingle as we walk in, but no one looks our way.

That's cool, sometimes they make you wait, just for good measure. That isn't so unusual, I think to myself.

But as the seconds turn to uncomfortable minutes, the strangest damn scene starts to unfold.

I mean, picture this…

We're standing in the entry of this tiny restaurant, where there are several tables for groups of two and four people scattered about the room. A lot of the tables are filled, but not all of them. There are plenty of empty tables for customers.

All the people in the restaurant are talking with each other, discussing their daily goings on, like it's any other morning. There are three groupings of small families, children in tow; in one corner I notice a middle-aged couple with a fussy baby, and at the table next to them, a family with two young children who are arguing. An elderly couple sits ignoring the ruckus the children are making, and go on quietly eating.

There are three servers bustling to and fro; one person who's looking around like he's in a supervisory position, possibly the host. I can see into the kitchen through one of those cut-out windows; the cooks and servers are yelling the orders back and forth, shoving the plates of steaming food onto the counter, to be snatched up by the servers waiting to take the orders to their customers.

The strange thing is, nobody turns around to look at us when we come in. And it isn't like only a few people are ignoring us. I mean, **none** of the patrons, none of servers, neither the host, nor the cooks who were facing our direction, look at us or acknowledge us in the least.

The restaurant is a small place, long, and not very wide. It's impossible that they can't see us standing here, but nobody turns around.

It's as if we have somehow stepped into an episode of "The Twilight Zone."

I go over the sequence of events in my mind once again.

The bell rang.

We walked in and nobody looked. We stand here looking foolish.

As the food servers who have their backs to us grab dishes of food and walk toward us to put plates down in front of other people waiting for their food, they pass right by us as if we're not here. We don't even seem to register in their presence.

It's just, weird. Damned weird. At first, we don't believe what's happening. We look at each other and are like "ewww, spooky."

But as we begin to comment on how rude it is to ignore customers, we see no reaction. Then, in louder and louder voices, we begin discussing our complete disappearance or apparent invisibility, or that maybe this is candid camera.

But, still no reaction…no reaction at all.

The host doesn't walk over to us and ask us to leave (at least we're used to that sort of treatment.) People didn't walk in behind us, so it was like we were stuck.

Ok now this is too fuck'n weird. We just wanted something

to eat, not a supernatural disappearance, especially on an empty stomach.

Either we're not here, or we're here and they're ignoring us. But how is that possible? How could all of them – cooks, busboys, food servers, host and every patron, including the children – how could they have planned this?

Whatever.

"Let's fuck'n split man!" I pipe up REAL LOUD.

Speed stomps out of the place. We all follow, taking an immediate left into the next little joint that was open. We aren't looking for five star quality service here folks. We just want to register on the scale of existence!

And we find a place, by God. The next place we walk into is way better.

It turns out, we actually do exist.

And we're famished.

We place our orders, get our food and sit at the nearest booth we can stumble into. As the restaurant begins to fill up with other patrons, we eat and talk about what just happened, not quite believing it ourselves.

From that day forth, whenever "the LA Four" got together, we'd still rehash "The Twilight Zone" incident.

Right up until the day Time Ripped, and tore us all apart forever.

Endings and Beginnings

Barrage noted the erratic readouts from the energy signatures in the atmosphere around his ship. He checked for any space going craft within a thousand parsecs and found, to his consternation, that a ship was following his (or least it was on the same trajectory as his, and that of the smaller craft he was following.)

Curious as to why the tiny, two manned craft was in the area, he ran full scans on their equipment, sure that he would find a Terraxian registry.

Not surprised that he was right, he felt a slight twinge of irritation that perhaps he'd been followed, however clumsy the attempt. Checking the data download from the craft (that was no bigger than a 'mop), he found a decided lack of weapon signatures, and ascertained that it was not, in fact, a MOB owned spacecraft.

So, who are these guys, and what do they hope to accomplish out here in the way beyond?

Now confident that it wasn't some kind of double cross from dear old dad, Barrage was free to ponder the seeming coincidence, and implications of three rather inadequately prepared spacecraft, all heading on the same course, straight away from Terrax.

Without a deep space port, surely their craft will run out of supplies and fuel within a few hundred parsecs.

Hmmph. I'll be damned if I'm going to share my supplies, or rescue these ill-prepared stragglers from themselves.

Wait a minute! That smaller craft is being piloted by my Senselle vixen Randi! What in the name of Asna's Eye is she doing out here? Are they following me after all?

Monitoring their incoming messages, he noted a lot of troubled confusion from the WC commercial craft attempting to approach the outer pre-docking orbits around Terrax.

Convincing himself he was only bored, and had no real interest in what was going on back on Terrax, he frowned slightly as he continued to switch from commercial channels to the usually orderly military comm bands.

The mil comm bands were alive with the staccato of mil speak; short bursts of acronyms and abbreviated phrases used in various combinations to form messages, but they all seemed to be saying the same thing.

It appeared that Terrax was under some sort of aggressive attack from an unknown source or sources. For no apparent reason, the energy signatures from the transceiving equipment on the Hives were receiving too much star energy/matter, and the input was increasing exponentially.

Energy was being created much faster than the transformers could handle it, threatening an overload in the system planetwide.

Of course, the military assumed it was an intergalactic attack coming from the Astral77 Universal Consortium, an old and wily enemy. However, the MOB had been tracking the A77UC's movements for some time, and Barrage was unaware that they had the power to cross the space bridge between their universal position and that of the Terraxian Entry Portal, as of yet.

No, whatever this was, it wasn't from a familiar foe.

Unable to communicate with the other vessels via traditional means, Barrage prepared to jack-in to his comm ports, and try to break through the babble as pure energy by transmitting himself off-ship via the local comm bands.

He fervently hoped he didn't miscalculate their energy sigs and accidentally project himself into the ether.

* * *

Cami was having a helluva time trying to monitor Wad Medani's MMs, and Auntie Vee's Panatechs were in a similar uproar. It seemed that the volume of frantic messaging was due to increased temperatures in the Hives and Warrens, along with the dangerously high energy inputs to the transformers, was causing the viewership to panic on a massive scale.

Cami desperately wished she could get through to her old Senselle colleague, Bepko da Silva, to see how the "Great Experiment" had worked out.

Did the crazy bastids finally succeed in disimaging and reimaging Auntie Vee and Wad Medani? Did her mentors finally consummate their long-lost love?

Cami was monitoring a 144-mini-screen projection percomp displaying internal MOB frequencies, public and local Senselle channels, general viewership feeds, input from Auntie Vee's labs and Wad Medani's MM boards, as well as downloads from Gramps and his old cronies.

Trying to make some sense out of the confusion, Cami suddenly became aware that the energy approaching Terrax was not concerned with matter at all, either military or mundane.

Cami recognized the energy signature of an impossibly

massive “edge of cloud” effect emanating from the borders of a dark energy mass which appeared to be heading directly toward the largest biophoto array on Terrax.

The magnificent Melissae Hive, jutting skyward and shining as if on fire, served as a celestial beacon. Gracing the coastline along the Northern OceanDome, which itself glittered with prized organic polycellular starmatter collectors, the combination proved irresistible to the energy consuming cloud of Phantom Energy approaching Terrax.

The light beckoned, as a flame to dark moths.

The energy cloud appeared to be swirling with movement; a mass of dark/light energy/matter containing more void than density.

She could hear from the ensuing chaos of communications by the viewership as the local comm bands were being disrupted. The satellites orbiting Terrax began to burn out, one by one. The light from the smaller Hives began to dim, as all non-essential energy was being utilized trying to intercept and repel the “edge-of-cloud” energy build-up.

The combined efforts of the MOB and WC and Senselle groups had little effect on the dark mass.

Slowly, with a palpable shock from the frightened viewership, the comm bands were silenced, as the lower frequencies were snapped off.

No more were the comfortably familiar volume of voice-mails, emails, IMs, PMs, peeps, RSS feeds, or streaming VR grammes.

The world glowed fiercely in eerie near silence.

Cami stopped monitoring the comm bands and let Randi take over for a while, her mind grinding to an uneasy halt. She didn’t know what could be done, and so sat with her nose pointed to a blank wall, trying to silence her mind and ignore the dread she felt building in the pit of her stomach.

Randi sat listening transfixed, as the world seemed to shrink to a pinpoint in Time.

***

When Frankie and Arpad came up for air after a three day debauch, they couldn't believe what they heard. The usually quiet deep-space, comm bands were registering two other ships in the area, both alive with incoming and outgoing requests for information and clarification of the status of their home world.

As if that weren't mystery enough, Frankie's old Senselle pals, Cami and Randi, were following their ship (for some unknown reason), and they couldn't get a word in edgewise, what with the increasingly urgent requests for permission to dock in Terraxian space.

Unfortunately, there was no response from the WC controlled Cheops Weather Service for clearance, leaving cargo barges crammed with interstellar trade goods cooling their jets alongside HORSEs with precious H^2O Retrieval tanks full, and the usual spate of pleasure cruisers, personal transports, galactic visitors, indignant World Council dignitaries, and military vessels all vying for permission to dock.

All were kept waiting silently.

Frankie and Arpad exchanged glances of concern and curiosity, respectively, and continued to listen to the mounting chaos, wondering what they could do for their helpless home world.

***

the Dark Ones come upon us

Light.

Matter.

We are drawn to the light/matter.

We come.

***

THE END

Rave Reviews

"My congratulations to Diane Tegarden for her book Anti-Vigilante and The Rips in Time. I found it to be a very entertaining read; superbly written with well-defined characters and an intriguing plot. Once picked up, it's tough to put down. Great job, Diane!"

- DENNY GRIFFIN, AUTHOR
HTTP://WWW.DENNISNGRIFFIN.COM

"As I writer I try to support my fellow literary engineers every time the opportunity presents itself. Good writers are the ones who keep you enthralled and turning the pages and this is exactly what happened to me when I read chapter one of Diane Tegarden's 'Anti-Vigilante and the Rips in Time'.

Chapter one of "Anti-Vigilante and the Rips in Time" is a delectable appetizer that makes you eager to delve into the main course.

Forget the New York Times list and read a really good book by a really creative writer. I give it ten thumbs up!!!"

- MARSHA, LAMBKIN WALTERS LAVENDER PUBLISHING
WWW.LAMBKINWALTERSLAVENDER.COM

"Diane Tegarden writes across genres. Her latest effort delves into planet and shows how society is changed by the use of renewable energy production. She delicately explains the beauty and folly of the human condition through some innovative ideas like a "re-imaging" machine that will enable people to relive emotions and experiences from their past.
Throughout the book, Tegarden interjects her native spirituality, detailing some of the rituals and dances of her native faith."

- PAM OSBEY, RADIO TALK SHOW HOSTESS
HTTP://WWW.BLOGTALKRADIO.COM/LITERARYPIZZAZZ

"ANTI-VIGILANTE AND THE RIPS IN TIME is a fascinating science fiction novel where even in a world where life is lived vicariously through virtual reality, murder is commonplace and carried out with extreme emphasis on brutality.

It never fails to amaze me -- the mind of the writers of science fiction, the terminology that is all their own, and a world created with such detail you think you are there.

This book is beautifully written about a place called Terrax (earth in the distant future). A world where human physical contact is impossible and pornography is an addiction, Murder a pastime and death just as horrible as it is in any reality.

For me the beauty of the book is the poetry and it is scattered very liberally throughout. All science fiction readers will enjoy this look at a world where the atmosphere is non-breathable and its population looks back in time at the mythology that is earth today.

A unique read by a very talented writer."

- YOLANDA RENEE, AUTHOR OF "MURDER, MADNESS & LOVE"
WWW.YOLANDARENEE.COM

"I've had the pleasure of readi ng the first chapter of Diane Tegarden's "Anti –Vigilante: Rips in Time". Diane has ability to manipulate words to create images showing the lives of her unique and interesting characters being interrupted by the insanity of The Milliner, a sadistic murderer that skins the flesh off his victims' faces. The visions she creates are both disturbing and intriguing at the same time. It is this technique in her writing skill that kept me on the edge of my seat reading to find out what happens next. I highly recommend reading Diane Tegarden's book "Anti-Vigilante: Rips in Time" and read for yourself!"

- KRISTINE SHEEHAN. OWNER
THE MERRY BIRD...PEN, INK AND DESIGN

Source Cited

Ocean Energy Recovery-The State of the Art, edited by Richard J. Seymour, published by the American Society of Civil Engineers, 345 East 47th Street, New York, N.Y. 10017

(1) Chapter 2, Warnock, J. Gavin and Clark, Robert H. page 17.

(2) Chapter 4, Baker, G.C. page 43.

(2a) Also: consulting engineer, 536 Main Street, Kentville, Nova Scotia, Canada, B4N 1L3. page 42.

(3) Chapter 8, Carmichael, A. Douglas and Falnes, Johannes. page 188.

About the Author

Diane Tegarden, an author of three published books, lives in Pasadena, with her husband, three cats and a dog.

She's been a print journalist since 1992, writing on a wide range of subjects including; renewable energy, environmental concerns, holistic health, women's issues, and Native Spirituality. Her work has been published in: The Pasadena Star News, Security Sales Magazine, The Pasadena Weekly, The Solar Flare, The Feral Forest, Earth Luvyrs Calendar, The Daily Sundial, the West Coast Well Being, Associated Content and Health Breakthroughs.

In April 2004, she self-published her first book, "Getting Out Of Limbo– A Self Help Divorce Book For Women," which provides step by step instructions on how to achieve financial and emotional independence in preparation for a divorce. Recently she's been interviewed on 20 Radio Talk Shows about divorce and domestic violence, referencing her experiences in "Getting Out of Limbo– A Self Help Divorce Book for Women" during Women's History Month, in March of 2008-2009.

Her company, FireWalker Publications Inc. was formed in 2004 to promote her passionate, original written works, materials, ideas and creations.

Since 2005, she's written for two online content producers, Associated Content and Constant Content.

She released her first poetry ebook, "Light Through Shuttered Window", in December 2007. The 90 page ebook is a compendium of her poems and songs about life, love and the process of creativity.

Her newest release "Anti-Vigilante and The Rips in Time" is a science fiction novel set in the distant future with its hooks deep into renewable energy and the changing face of the planet.

To Contact Ms. Tegarden, please go to her website at: www.firewalkerpublications.com

Breinigsville, PA USA
01 June 2010
239043BV00001B/13/P

9 780974 536910